Advances in Synthesis, Characterization, and Application of Thin Films

Advances in Synthesis, Characterization, and Application of Thin Films

Guest Editors

Miłosz Grodzicki
Damian Wojcieszak
Michał Mazur

MDPI

Basel • Beijing • Wuhan • Barcelona • Belgrade • Novi Sad • Cluj • Manchester

Guest Editors
Miłosz Grodzicki
Department of Semiconductor Materials Engineering
Faculty of Fundamental Problems of Technology
Wrocław University of Science and Technology
Wrocław
Poland

Damian Wojcieszak
Department of Electronic and Photonic Metrology
Faculty of Electronics, Photonics and Microsystems
Wrocław University of Science and Technology
Wrocław
Poland

Michał Mazur
Department of Electronic and Photonic Metrology
Faculty of Electronics, Photonics and Microsystems
Wrocław University of Science and Technology
Wrocław
Poland

Editorial Office
MDPI AG
Grosspeteranlage 5
4052 Basel, Switzerland

This is a reprint of the Special Issue, published open access by the journal *Crystals* (ISSN 2073-4352), freely accessible at: https://www.mdpi.com/journal/crystals/special_issues/5VA621SX6D.

For citation purposes, cite each article independently as indicated on the article page online and as indicated below:

Lastname, A.A.; Lastname, B.B. Article Title. *Journal Name* **Year**, *Volume Number*, Page Range.

ISBN 978-3-7258-2971-2 (Hbk)
ISBN 978-3-7258-2972-9 (PDF)
https://doi.org/10.3390/books978-3-7258-2972-9

Contents

Preface . . . . **vii**

Shuai Wu, Kesheng Guo, Jie Bai, Jiafeng Li, Jingming Zhu, Lei Liu, et al.
Study on the Simulation and Experimental Impact of Substrate Holder Design on 3-Inch High-Quality Polycrystalline Diamond Thin Film Growth in a 2.45 GHz Resonant Cavity MPCVD
Reprinted from: *Crystals* **2024**, *14*, 821, https://doi.org/10.3390/cryst14090821 . . . . **1**

Thom R. Harris-Lee, Andrew Brookes, Jie Zhang, Cameron L. Bentley, Frank Marken and Andrew L. Johnson
Plasma-Enhanced Atomic Layer Deposition of Hematite for Photoelectrochemical Water Splitting Applications
Reprinted from: *Crystals* **2024**, *14*, 723, https://doi.org/10.3390/cryst14080723 . . . . **16**

Erick Gastellóu, Rafael García, Ana M. Herrera, Antonio Ramos, Godofredo García, Gustavo A. Hirata, et al.
Deposition and Structural Characterization of Mg-Zn Co-Doped GaN Films by Radio-Frequency Magnetron Sputtering in a N_2-Ar_2 Environment
Reprinted from: *Crystals* **2024**, *14*, 618, https://doi.org/10.3390/cryst14070618 . . . . **33**

Yugang Li, Zhiyong Liu, Ping Zhu, Jinyu He and Chuanbing Cai
High-Temperature (Cu,C)$Ba_2Ca_3Cu_4O_y$ Superconducting Films with Large Irreversible Fields Grown on $SrLaAlO_4$ Substrates by Pulsed Laser Deposition
Reprinted from: *Crystals* **2024**, *14*, 514, https://doi.org/10.3390/cryst14060514 . . . . **43**

Yifei Li, Tiwei Chen, Yongjian Ma, Yu Hu, Li Zhang, Xiaodong Zhang, et al.
Ultrahigh Responsivity In_2O_3 UVA Photodetector through Modulation of Trimethylindium Flow Rate
Reprinted from: *Crystals* **2024**, *14*, 494, https://doi.org/10.3390/cryst14060494 . . . . **52**

Marcos Palacios Bonilla, Godofredo García Salgado, Antonio Coyopol Solís, Román Romano Trujillo, Fabiola Gabriela Nieto Caballero, Enrique Rosendo Andrés, et al.
Influence of the Incorporation of Nd in ZnO Films Grown by the HFCVD Technique to Enhance Photoluminiscence Due to Defects
Reprinted from: *Crystals* **2024**, *14*, 491, https://doi.org/10.3390/cryst14060491 . . . . **64**

Dauren B. Kadyrzhanov, Medet T. Idinov, Dmitriy I. Shlimas and Artem L. Kozlovskiy
The Influence of Variations in Synthesis Conditions on the Phase Composition, Strength and Shielding Characteristics of $CuBi_2O_4$ Films
Reprinted from: *Crystals* **2024**, *14*, 453, https://doi.org/10.3390/cryst14050453 . . . . **75**

Chao Li, Wenxin Li, Guangqin Wu, Guojin Chen, Junyi Wu, Niushan Zhang, et al.
Design and Study of Composite Film Preparation Platform
Reprinted from: *Crystals* **2024**, *14*, 389, https://doi.org/10.3390/cryst14050389 . . . . **102**

Katarzyna Lament, Miłosz Grodzicki, Radosław Wasielewski, Piotr Mazur and Antoni Ciszewski
Growth and Properties of Ultra-Thin PTCDI-C8 Films on GaN(0001)
Reprinted from: *Crystals* **2024**, *14*, 201, https://doi.org/10.3390/cryst14030201 . . . . **117**

Tatyana Ivanova, Antoaneta Harizanova, Tatyana Koutzarova, Benedicte Vertruyen and Raphael Closset
Sol–Gel Synthesis of ZnO:Li Thin Films: Impact of Annealing on Structural and Optical Properties
Reprinted from: *Crystals* **2024**, *14*, 6, https://doi.org/10.3390/cryst14010006 . . . . **130**

Preface

Thin films play a key role in modern science and technology, combining fundamental research and industrial applications. Their diverse structural, optical, and electronic properties make them essential components in various fields, such as optoelectronics, sensors, and energy conversion. Rapid advances in thin film synthesis methods combined with novel characterization techniques have significantly expanded their potential applications and opened up new avenues for innovation.

This Special Issue, "Advances in Synthesis, Characterization, and Application of Thin Films," features 10 carefully selected original research papers, with a primary focus on advancing the synthesis of various materials in the form of thin films.

We are confident that the articles in this Special Issue will inspire researchers to delve deeper into the science of thin films, contributing further innovation and opportunities for collaboration across disciplines. This collection exhibits the potential of thin film in shaping future scientific and industrial areas.

As Guest Editors of this Special Issue, we wish to express our gratitude to the contributing authors and peer reviewers. We also gratefully acknowledge MDPI's editorial team for their efforts in making this Special Issue a success.

Miłosz Grodzicki, Damian Wojcieszak, and Michał Mazur

Guest Editors

crystals

MDPI

Article

Study on the Simulation and Experimental Impact of Substrate Holder Design on 3-Inch High-Quality Polycrystalline Diamond Thin Film Growth in a 2.45 GHz Resonant Cavity MPCVD

Shuai Wu [1], Kesheng Guo [2,*], Jie Bai [3], Jiafeng Li [2], Jingming Zhu [2], Lei Liu [3], Lei Huang [2], Chuandong Zhang [3] and Qiang Wang [1,3,*]

1 School of Materials Science and Engineering, Chongqing Jiaotong University, Chongqing 400074, China
2 Jihua Laboratory, Foshan 528200, China
3 School of Aeronautics, Chongqing Jiaotong University, Chongqing 400074, China
* Correspondence: guoks@jihualab.com (K.G.); wangq@cqjtu.edu.cn (Q.W.)

Abstract: In this study, three different substrate holder shapes—trapezoidal, circular frustum, and adjustable cyclic—were designed and optimized to enhance the quality of polycrystalline diamond films grown using microwave plasma chemical vapor deposition (MPCVD). Simulation results indicate that altering the shape of the substrate holder leads to a uniform distribution of the electric field on the surface, significantly suppressing the formation of secondary plasma. This design ensures a more even distribution of the temperature field and plasma environment on the substrate holder, resulting in a heart-shaped distribution. Polycrystalline diamond films were synthesized under these three different substrate holder conditions, and their morphology and crystal quality were characterized using optical microscopy, Raman spectroscopy, and high-resolution X-ray diffraction. Under conditions of 5 kW power and 90 Torr pressure, the adjustable cyclic substrate holder produced high-quality 3-inch diamond films with low stress and narrow Raman full width at half maximum (FWHM). The results confirm the reliability of the simulations and the effectiveness of the adjustable cyclic substrate holder. This approach provides a viable method for scaling up the size and improving the quality of polycrystalline diamond films for future applications.

Keywords: 2.45 GHz; MPCVD; polycrystalline diamond thin film; design of substrate holder; numerical simulation

Citation: Wu, S.; Guo, K.; Bai, J.; Li, J.; Zhu, J.; Liu, L.; Huang, L.; Zhang, C.; Wang, Q. Study on the Simulation and Experimental Impact of Substrate Holder Design on 3-Inch High-Quality Polycrystalline Diamond Thin Film Growth in a 2.45 GHz Resonant Cavity MPCVD. *Crystals* **2024**, *14*, 821. https://doi.org/10.3390/cryst14090821

Academic Editor: Dah-Shyang Tsai

Received: 19 August 2024
Revised: 10 September 2024
Accepted: 14 September 2024
Published: 20 September 2024

1. Introduction

The development of wide-bandgap semiconductor materials such as SiC [1], GaN [2], and AlN [3] has revolutionized various fields, including microelectromechanical systems, aerospace materials, information sensing, acoustic filters, and quantum technology [4,5]. These materials are favored because of their high breakdown electric fields, electron mobility, and excellent environmental stability. However, operating at higher power levels presents significant challenges owing to heat-dissipation issues, which limit performance and compromise device reliability. Effective thermal management is essential for addressing this challenge. Diamond has emerged as a highly promising material for thermal management because of its extremely high thermal conductivity (2000 W/mK), high radiation hardness, and excellent chemical stability [6]. Integrating diamond with wide-bandgap semiconductors can enhance heat dissipation, thereby increasing the area-dissipation density and reducing the working-channel temperature. This integration offers a viable solution to improve the performance and reliability of devices operating at high power levels [7].

Various chemical vapor deposition (CVD) techniques have been employed to deposit large-area, high-quality polycrystalline diamond films. These high-quality films are generally characterized by a low residual stress, fewer defect peaks, absence of impurity phases, and high crystallinity. This process typically involves the reaction of a mixture of gases (e.g., CH_4 and H_2) on a substrate, resulting in the formation of polycrystalline diamond films [8,9]. Currently, MPCVD is the most refined method for producing synthetic diamonds. A mixture of gases (hydrogen and carbon source gases) is introduced, and the microwaves generated by the microwave source are transmitted through a rectangular waveguide. After passing through a mode converter, the microwaves are coupled to the resonant cavity. A microwave resonator creates a strong and uniform standing-wave electric field that ionizes gases to form a plasma ball [10]. The substrate is placed beneath the plasma ball for the diamond film growth. This electrode-free process eliminates potential contamination sources, enabling the deposition of high-quality polycrystalline diamond films on the substrate. The resonant cavity in MPCVD equipment is characterized by microwave resonance. When microwaves of the resonant frequency are input into the cavity, resonance occurs, creating a high-intensity electric field. This electric field can be adjusted by modifying the size and shape of the cavity to achieve an optimal electric field distribution and generate a plasma ball at a specific location. However, this requires high precision in cavity machining because the presence of plasma also affects the resonant cavity. Therefore, to optimize the design of the resonant cavity, it is necessary to simulate the electric field and calculate the plasma distribution [11,12].

The process of growing high-quality polycrystalline diamond thin films is influenced by numerous parameters, including substrate selection [13], microwave power [14], chamber pressure [15], gas composition [16], chamber temperature [17], plasma ball density [18], and substrate support [19]. Additionally, the cavity size and frequency of MPCVD equipment play significant roles. Currently, the MPCVD frequencies used are 915 MHz and 2.45 GHz. When comparing microwaves of two different frequencies, the 915 MHz frequency has a longer wavelength, resulting in a larger plasma ball. Consequently, the required microwave power is higher, leading to a larger diameter of deposited diamond film. However, MPCVD equipment operating at 915 MHz demands higher power, making the structure and details of the equipment more complex and posing significant challenges for maintaining vacuum conditions [20,21]. This, in turn, increases the requirements for equipment development, manufacturing technology, and cost. On the other hand, MPCVD operating at 2.45 GHz can achieve high plasma density and deposition rates and has been widely adopted. In general, when microwaves are coupled to a resonant cavity, two resonance modes exist: transverse electric (TE) and transverse magnetic (TM). The inner metal walls of the cavity enforce a zero-tangential component on the electric field, causing it to be perpendicular to the inner wall surface. In the TE mode, there are no strong electric field regions in contact with the metal, meaning the plasma cannot contact the substrate surface. Therefore, the TM mode is commonly used, with TM_{0mn} being the most prevalent (where 0 indicates an axially symmetric surface electric field structure, and m and n represent the number of axial and radial electric field maxima in the resonant cavity, respectively). Moreover, the substrate holder within the cavity plays a crucial role in the effective utilization of microwave energy [22–24]. To ensure maximum energy utilization, a substrate holder is added to the reaction zone to focus the electric field above the substrate, thereby avoiding the formation of secondary electric fields. In 2017, An et al. [25] observed that reducing the substrate support size radially and increasing it axially resulted in an uneven plasma distribution. They discovered that altering the height of the movable substrate could achieve uniform plasma and power density. In 2022, a T-shaped substrate was developed, introducing a gap at the edge that created a strong additional electric field, enhancing the edge temperature uniformity and improving the quality of the deposited polycrystalline diamond films. Zhao et al. [26] simulated and introduced a hole in the center of the substrate stage and found that it suppressed nitrogen gas in the cavity without significantly affecting the plasma, although the electron density decreased by 40%. Sedov et al. [19] designed three different geometric shapes for substrate platforms using electric field simulations and manufactured two-inch polycrystalline diamond thin films through simulation testing. However, these platforms exhibited strong edge effects.

In this study, a butterfly-shaped resonant cavity was employed within a frequency-domain transient solution mode, coupled with multiple physical fields. Various substrate holder designs were proposed to investigate their impact on the microwave electric field, plasma environment, and the quality of diamond films. Numerical simulations and experimental tests were conducted to identify the optimal substrate holder for the deposition of high-quality diamond.

2. Simulation and Experiment

2.1. Simulation Modeling

Figure 1a shows a schematic of the butterfly resonant cavity MPCVD used in the experiment. Using quartz as the dielectric window, this setup can be classified as a quartz-ring MPCVD, which is a non-cylindrical plasma device. Microwaves are input from below, coupled into the stainless-steel reaction chamber through a quartz window, and focused above the substrate stage to form a strong electric field. At the bottom of the cavity, there is a single substrate holder with a width of 220 mm for placement of the substrate. The remaining dimensions are shown in Figure 1a. The electric field distribution in Figure 1b shows a strong secondary electric field near the top of the cavity. At high power, this can lead to the formation of secondary plasma, resulting in capability loss.

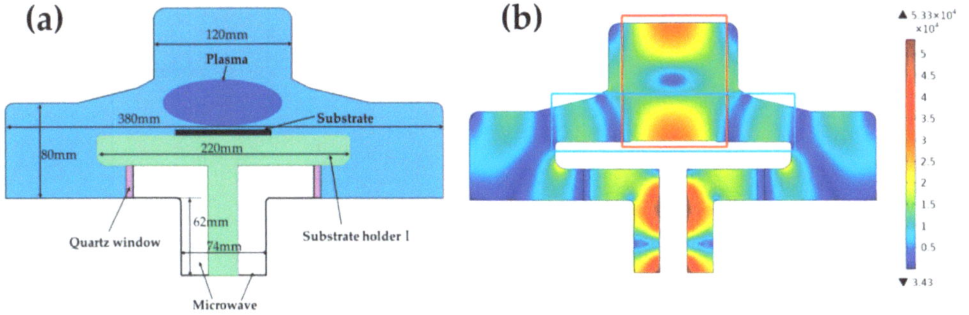

Figure 1. (**a**) Schematic diagram of a 2.45 GHz butterfly resonant cavity MPCVD and (**b**) electric field distribution without plasma generation at an input power of 1.5 kW.

To address the issue of secondary plasma formation and achieve high-quality polycrystalline diamond films, we designed three different substrate holders to suppress secondary plasma. In addition, we developed a substrate platform that ensures a highly uniform plasma field distribution. Figure 2 shows a schematic diagram of the three substrate holders, all made of Mo. These holders were designed to tune microwaves, adjust electric fields, and optimize plasma and temperature distribution.

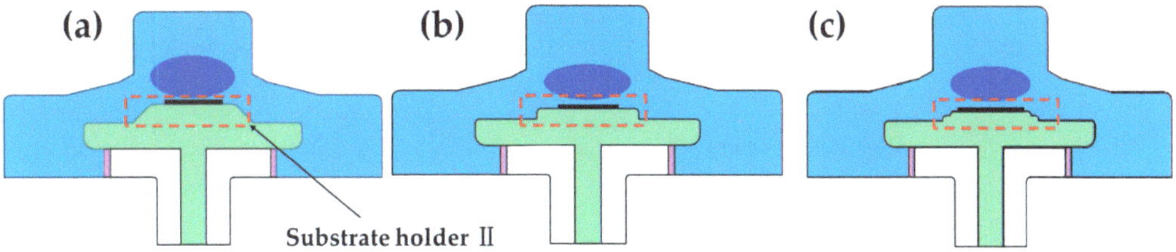

Figure 2. The three designed substrate holders display (**a**) trapezoidal, (**b**) circular frustum, and (**c**) adjustable cyclic shapes in the red box.

To simulate the experimental process, a finite element analysis method was used to study the electric field and plasma electron density distribution inside the cavity. During the simulation, we used a microwave power of 1000 W, working pressure of 40 Torr, and microwave frequency of 2.45 GHz. The plasma model approximates the Boltzmann equation by using a fluid equation. The rate of change in the electron density can be described by Equation (1):

$$\frac{\partial}{\partial t}(n_e) = R_e + \nabla \cdot \Gamma_e \tag{1}$$

In the formula, Γ_e is the electron flux, and R_e is the electron source.

The definition of electron flux Γ_e is as follows:

$$\Gamma_e = \mu_e n_e E + \nabla(D_e n_e) \tag{2}$$

The electronic source R_e is defined as

$$R_e = \sum_{j=1}^{M} x_j k_j N_n n_e \tag{3}$$

where x_j is the molar fraction of the target substance for reaction j, k_j is the rate constant for reaction j, and N_n is the total number density of neutral particles.

The rate of change of electron energy density is

$$\frac{\partial}{\partial t}(n_\varepsilon) = R_\varepsilon + \nabla \cdot \Gamma_\varepsilon + E \cdot \Gamma_e \tag{4}$$

where n_ε is the electron energy density, and R_ε is the loss or increase in ability caused by inelastic collisions. Γ_ε is defined as the electron energy flux:

$$\Gamma_\varepsilon = \mu_\varepsilon n_\varepsilon E + \nabla(D_\varepsilon n_\varepsilon) \tag{5}$$

The microwave electric field distribution of the MPCVD device is solved by Maxwell's equation:

$$\nabla \times \mu_r^{-1}(\nabla \times E) - k_0^2\left(\epsilon_r - \frac{j\sigma}{\omega \epsilon_0}\right)E = 0 \tag{6}$$

In the formula, E is the electric field, ω and k_0 are the angular frequency and wavenumber of microwaves, ϵ_0 is the vacuum dielectric constant, μ_r, ϵ_r, and σ are the relative magnetic permeability, relative dielectric constant, and conductivity of the material, respectively, and j is the imaginary unit. In the non-discharged gas region, $\epsilon_r = 1$, $\sigma = 0$; in the quartz glass region, $\epsilon_r = 3.78$, $\sigma = 0$; and for the discharged gas region, the conductivity is given by the following formula:

$$\sigma = \frac{n_e q^2}{m_e(\nu_e + j\omega)} \tag{7}$$

In the formula, q and m_e are the charge and mass of electrons, respectively, n_e is the electron density, and ν_e is the electron neutral particle collision frequency of the plasma.

During the diamond deposition process, the introduction of a small amount of methane does not have a significant impact. Therefore, only the discharge process of H_2 is considered. The hydrogen plasma reaction was studied based on the work of K. Hassouni [27]. Our model includes only e, H_2, H, H (n = 2), H (n = 3), H_+, and H_2^+. In the calculations, the gas temperature is set to be equal to the ion temperature. The following are some important H_2 plasma reactions:

$$e + H_2 \rightarrow 2e + H_2^+ \tag{R1}$$

$$e + H \rightarrow e + H^* \tag{R2}$$

$$e + H_2 \rightarrow e + 2H \tag{R3}$$

$$e + H_2 \rightarrow e + H_2 \tag{R4}$$

$$H_2^+ + H \rightarrow H^+ + H_2 \tag{R5}$$

$$H^{**} \rightarrow H + hv \tag{R6}$$

The electronic energy loss R_ϵ is obtained by summing up the collision energy losses of all reactions:

$$R_\epsilon = \sum_{j=1}^{p} x_j k_j N_n n_e \Delta\epsilon_j \tag{8}$$

In the equation, $\Delta\epsilon_j$ is the energy loss of reaction j. The reaction rate was calculated using reaction collision cross-section data and the electron energy distribution function:

$$k_j = \gamma \int_0^\infty \epsilon \sigma_k(\epsilon) f(\epsilon) d\epsilon \tag{9}$$

Among them , $\gamma = \sqrt{2q/m}$, $\sigma_k(\epsilon)$ is the collision cross-section, and $f(\epsilon)$ is the electron energy distribution function (EEDF).

In addition, the simulation of MPCVD equipment must also consider the impact of gas temperature, assuming the effects of convection are negligible:

$$\nabla\cdot(-k\nabla T_g) = Q \tag{10}$$

In the equation, k represents the thermal conductivity of the gas, T_g is the gas temperature, and Q denotes the heat source of the gas. In the MPCVD device, the average free path of electrons is very short, and the ability of electrons to absorb from microwaves is transferred to the gas in a short period of time, causing the gas temperature to rise. Therefore, when calculating, the heat source Q can be approximated as the power density Q_h.

2.2. Experimental Details

Polycrystalline diamond thin films were deposited on a 3-inch monocrystalline silicon (100) substrate. Before diamond deposition, the Si substrate was immersed in a nanodiamond suspension and sonicated for 30 min to enhance the diamond nucleation capability. After drying, the substrates were prepared for diamond growth. The processed silicon substrate was then placed on three differently designed deposition supports in alignment with the simulated configurations. The growth conditions were set as follows: microwave power of 5 kW, pressure of 90 Torr, and H_2/CH_4 flow rates of 300 and 9 sccm, respectively. By comparing the properties of the grown polycrystalline diamond films with the simulation results, the optimal substrate support for achieving high-quality polycrystalline diamond films was identified.

An infrared thermometer was used to measure the temperatures of the substrate support and Si during the deposition process through a window at the top of the cavity. High-resolution field-emission scanning electron microscopy (FE-SEM; Verios 5 UC, Thermo Fisher, Brno, Czech Republic) was used to examine the surface morphology of the polycrystalline diamond films. Additionally, a confocal laser Raman microscope (Renishaw in Via Qontor, argon ion laser, London, UK, 514.5 nm, 50 mW, spectral resolution ≤ 1 cm^{-1}) was utilized to analyze the structural information of defects and deposits within the polycrystalline diamond films.

3. Results and Discussion

3.1. Influence of the Substrate Holder on the Electric Field in the Cavity

First, through simulation, we calculated the electric field distribution inside the cavity for different substrate holders. Studying the influence of electric fields before plasma generation is crucial. As shown in Figure 3a, two strong electric field regions appear in the cavity: one at the top of substrate holder I and the other at the top of the cavity. In this scenario, the electric field strength is very low and has a wide distribution range, which is extremely unfavorable for plasma generation. The presence of two electric fields within the cavity causes energy loss, resulting in an ineffective utilization of the input power for diamond deposition.

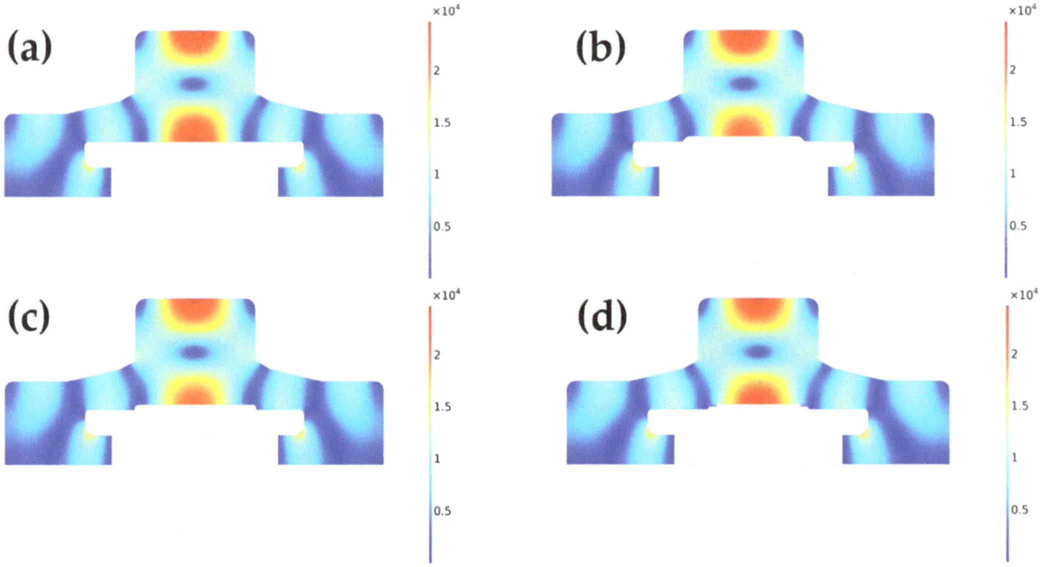

Figure 3. Electric field distribution diagram (V/m) of the MPCVD device with complete TM_{011} and TM_{021} modes, (**a**) no substrate holder II, (**b**) trapezoidal, (**c**) circular frustum, and (**d**) adjustable cyclic shape.

To enhance electric field utilization and reduce energy loss, Li et al. [28] added a splitter at the top of the cavity to shield the electric field focusing at the top, retaining only the strong electric field above the substrate stage inside the cavity. However, this design does not allow real-time adjustment of the substrate stage position, resulting in poor coordination. To address this issue, Zhang et al. [29] designed a cylindrical cavity by inverting it by 180°, which shields the electric field area above and concentrates the strong electric fields below the cavity for substrate placement. By optimizing the cavity parameters, they achieved a single high-density plasma sphere; however, this design only allowed the preparation of two-inch high-quality polycrystalline diamond films.

Compared with traditional cylindrical cavities, the butterfly resonant cavity in this study can generate a larger electric field and plasma size, which is beneficial for the large-scale and high-quality deposition of diamonds. To suppress the secondary electric fields, we propose three different substrate holders; their electric field distributions are shown in Figure 3. Previous studies have shown that placing a sample in a pocket-like substrate holder can protect the edges of the sample but results in an uneven electric field distribution [19]. Another study [30] suggested that increasing the height of the pocket substrate holder can improve the electric field, thereby enhancing the quality of the polycrystalline diamond films. Conversely, taller substrate holders demonstrate better electric field uniformity. In this study, we propose a trapezoidal substrate holder as shown in Figure 3b. The dimensions are 122 mm at the bottom, 5 mm in height, and 78 mm at the top with chamfered edges. A significant increase in the electric field intensity was observed, reaching 3.4×10^4 V/m. The electric field at the upper bottom of the trapezoidal holder weakened, and the strongest lateral electric field intensity was observed at the top of the cavity. Additionally, a strong edge effect was generated at the lower end of the original substrate holder, causing damage to the holder and resulting in electric field energy loss.

To address these issues, we propose a second type of substrate holder, II, designed as a circular frustum with a bottom diameter of 100 mm, an upper diameter of 90 mm, and a height of 5 mm, featuring chamfered edges. As shown in Figure 3c, this design reduces the edge effect observed at the lower end of the initial substrate holder, achieving an electric field intensity on the circular frustum of up to 4.4×10^4 V/m. Despite this improvement, the design exhibits an uneven electric field and lacks the ability to move vertically, which is crucial for adjusting the geometric shape of the resonant cavity and achieving optimal coordination. To enhance the design further, a lifting cylinder was added to the middle of the substrate holder. Figure 3d illustrates this improved design, which includes a circular platform with an upper diameter of 88 mm, a lower diameter of 100 mm, and a height of 5 mm. The movable cylinder, with a diameter of 78 mm, can be adjusted within an 8 mm range. This adjustment capability allowed for the coordination of the electric field during the growth process, resulting in a uniform electric field with a maximum intensity of 1.5×10^5 V/m.

3.2. *Influence of the Substrate Holder on Plasma Electron Density in the Cavity*

To better understand the influence of the substrate structure on the preparation of high-quality polycrystalline diamond films, it is essential to calculate not only the electric field distribution within the cavity but also the plasma electron density distribution. The principle of the resonant cavity involves focusing the electric field to excite the precursor gases, thereby forming a plasma [31]. In addition, the presence of plasma affects the electric field distribution, making the simulation of the plasma electron density distribution crucial [32]. A finite element method was employed to simulate plasma behavior in the butterfly cavity. To obtain reliable results, a simulation time of 2 s was selected for the plasma model calculations. This duration ensured the convergence of the model, providing accurate and consistent plasma electron density distribution data. The results of these simulations are essential for optimizing the substrate holder design, ultimately leading to the production of high-quality polycrystalline diamond films.

Figure 4 shows the plasma electron density distribution maps ($1/m^3$) for four different substrate holders. During operation of the MPCVD cavity, the electric field generates a secondary plasma electron located at the top of the cavity. This secondary plasma electron produces a shielding effect that weakens the plasma electrons above the substrate, hindering the high-quality deposition of polycrystalline diamond films [33]. The plasma electron density distribution is crucial in MPCVD because it directly affects the uniformity and rate of the deposition process. Figure 4a shows the plasma electron density distribution without the substrate holder II. A high-density plasma region is generated at the center of the substrate holder, with a maximum of 7.41×10^{17} m^{-3}, and the density decreases outward from the center. Under these conditions, the deposition rate at the center of the substrate was high, whereas that at the edges it was low, resulting in inconsistent and uneven diamond growth rates. Additionally, the heating temperature of the substrate inside the cavity is entirely from the plasma electrons and is too low for the optimal growth of polycrystalline diamonds. Figure 4b shows the plasma electron density distribution of the trapezoidal substrate holder II. The plasma electron density on the substrate holder decreases significantly, but a smaller plasma region is generated below the substrate holder. The secondary plasma electron density at the top of the cavity increases, which is unfavorable for the growth of polycrystalline diamond [34]. Figure 4c shows the plasma electron density distribution with the circular frustum substrate holder II. The high-density plasma region on the substrate holder extends to the left and right, changing the shape of the plasma sphere from circular to elliptical. This alteration may result in a more uniform deposition of polycrystalline diamond. However, it cannot be adjusted during the diamond growth process and requires further optimization. Figure 4d illustrates the plasma distribution with the adjustable cyclic-shaped substrate holder II. This holder retains the advantages shown in Figure 4c and exhibits a uniform and high-density area, facilitating the uniform deposition of polycrystalline diamond.

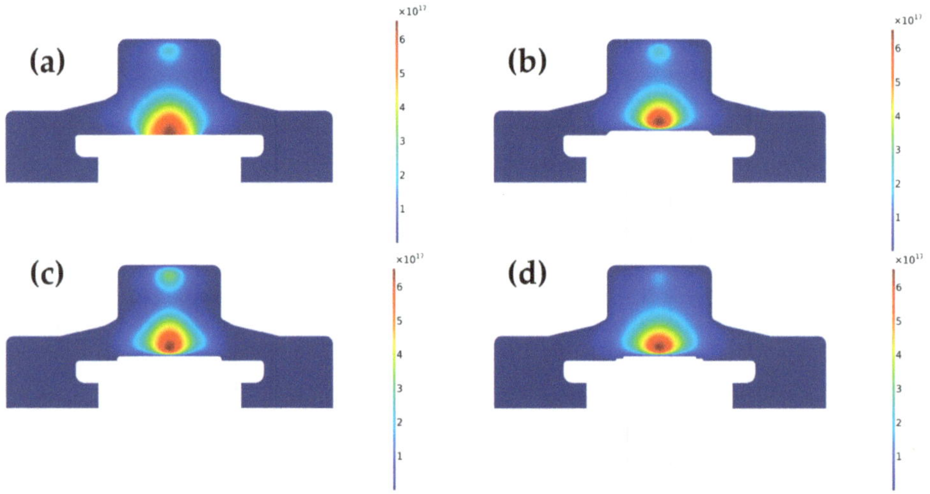

Figure 4. MPCVD equipment plasma electron density distribution map ($1/m^3$): (**a**) without substrate II, (**b**) trapezoidal, (**c**) circular frustum, and (**d**) adjustable cyclic shape.

3.3. Influence of the Substrate Holder on the Temperature of the Gas in the Cavity

Figure 5 shows the gas temperature distribution for different substrate holders. By comparing it with Figure 4, it is evident that gas temperature distribution correlates with plasma electron density distribution. As shown in Figure 5a, in the absence of substrate support II, a hat-shaped high-temperature zone will form in the plasma region. When the trapezoidal substrate holders are added, the high-temperature region becomes more dispersed, as seen in Figure 5b. In Figure 5c, the high-temperature region is slightly lower and spreads more towards the edges of the substrate holder. Figure 5d shows the broadest high-temperature region, indicating that the substrate holder design can maximize the area exposed to high temperatures, thereby improving the uniformity of diamond film deposition.

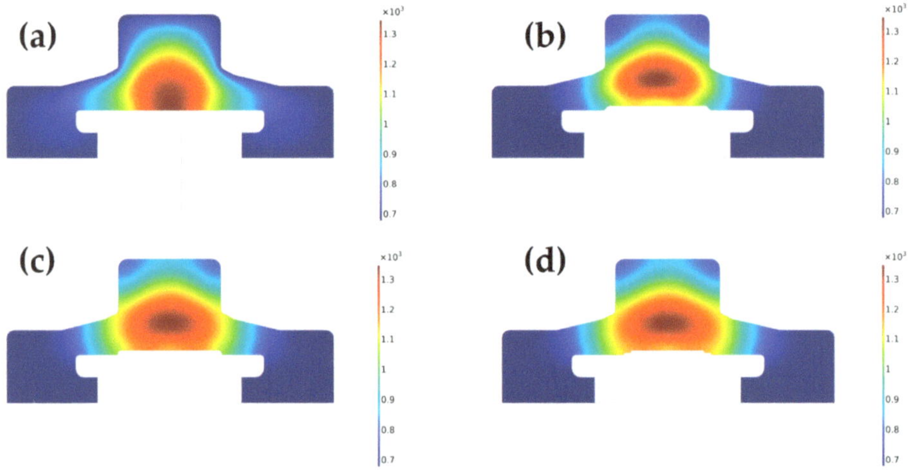

Figure 5. Temperature (K) distribution of plasma gas in MPCVD equipment: (**a**) without substrate II, (**b**) trapezoidal, (**c**) circular frustum, and (**d**) adjustable cyclic shape.

Figure 6 illustrates the number density distribution of H species along the central axis of the chamber under three different substrate holder configurations. The x-axis represents the height from the substrate surface. The figure reveals that the trend of the distribution curve varies with different substrate holders. Without a substrate holder, the number density generally increases with height. When a trapezoidal substrate holder is used, the H species number density is higher near the substrate surface and gradually decreases with increasing distance. In contrast, with the adjustable cyclic-shaped substrate holder, a uniform distribution of H species is observed within 10 mm from the substrate surface, and the overall trend shows a gradual decrease beyond this point. This uniform distribution supports the reliability of the proposed substrate holder design for producing high-quality diamond films.

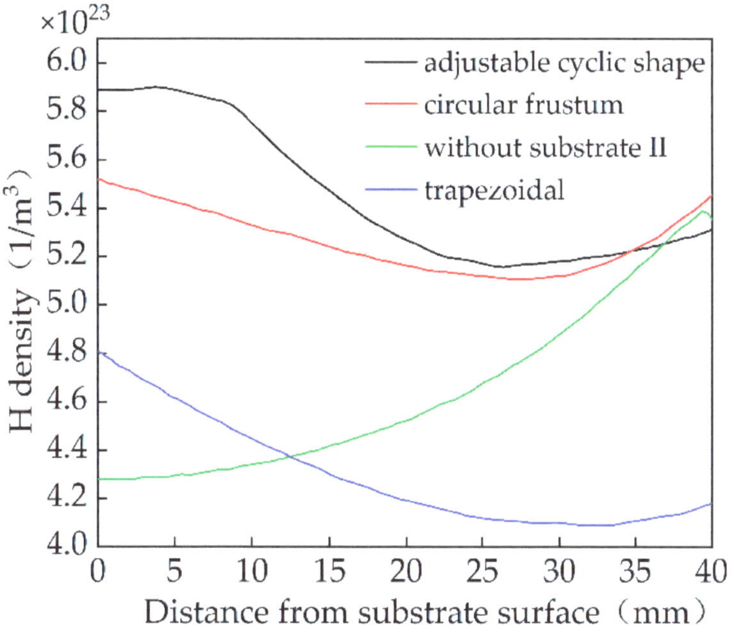

Figure 6. The number density distribution curve of substance H along the axis direction of the cavity.

3.4. Experimental Verification of Simulation Results

The actual MPCVD growth process is complex and requires experimental verification for an accurate simulation. Based on the simulation results, we manufactured three types of substrate holder II and placed them in a cavity for experimental testing. As shown in Figure 7, the actual plasmonic spheres generated using different substrate holders exhibited varying coverage and shapes. Figure 7a illustrates the trapezoidal substrate holder, in which the plasmonic sphere appears as a slightly curved elliptical shape with incomplete coverage. It is mainly concentrated in the center of the cavity with weak edges. Figure 7b shows the circular frustum substrate holder, where the plasma shape becomes more uniform, covering almost the entire top and bottom, and the brightness is more consistent. Figure 7c shows the adjustable cyclic shape, where the plasma appears flat and completely covers the substrate holder. These observations indicate that different substrate holders affect the plasma distribution and shape. Thus, optimization of the substrate holders can achieve improved plasma distribution and deposition effects. Additionally, it can be seen that the experimental plasma distribution results are largely consistent with the simulations.

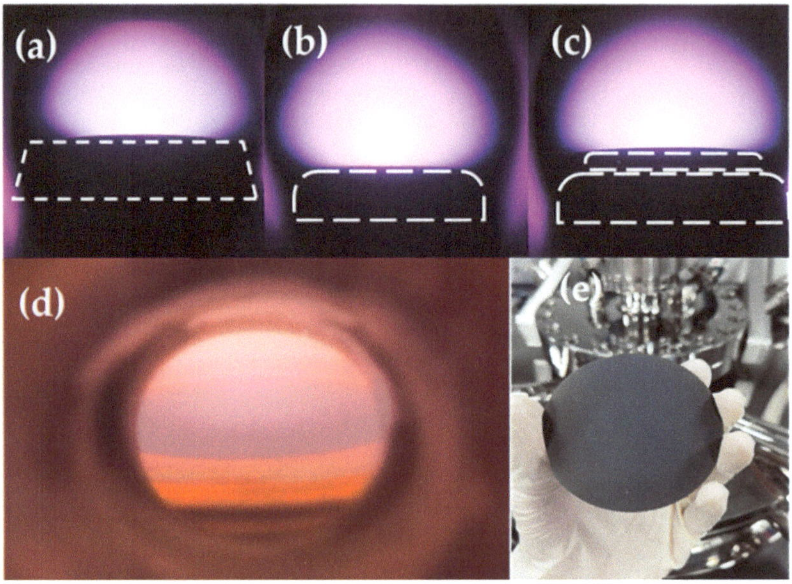

Figure 7. Plasma distribution diagram of the MPCVD device in a hydrogen atmosphere experiment: (**a**) trapezoidal, (**b**) circular frost, and (**c**) adjustable cyclic shape. (**d**) An actual image of polycrystalline diamond growth under an adjustable cyclic-shaped substrate holder. (**e**) Three inches of high-quality polycrystalline diamond growth under an adjustable cyclic-shaped substrate holder.

To verify the influence of the substrate holders, we used three types to analyze polycrystalline diamond films grown under identical parameters for four hours. The growth process using an adjustable cyclic substrate holder is illustrated in Figure 7d. The substrate holder turned bright red with plasma heating. Figure 7e shows the polycrystalline diamond thin film deposited in an adjustable cyclic-shaped substrate holder, which appears uniformly black. For comparison, we examined polycrystalline diamond films prepared using two other methods. Figure 8 shows the SEM images of these films. Figure 8a shows the results obtained using the trapezoidal substrate holder. The surface is not fully covered with polycrystalline diamond, displaying a graphite phase and 10-μm-sized polygonal diamond particles. Figure 8b shows complete coverage with a relatively small amount of graphite phase and a consistent structure within the region. Figure 8c shows that nearly full coverage of the surface was achieved. The polycrystalline diamond particles exhibited a uniform quadrilateral morphology with a lower growth rate than that shown in Figure 8b. This finding is consistent with those of other studies, suggesting that an excessively fast growth rate can reduce quality. The particle size distribution histograms indicate that the average particle sizes for the trapezoidal substrate holder, circular frustum, and adjustable cyclic shape are 4.24 μm, 4.49 μm, and 4.33 μm, respectively. The circular frustum exhibits the fastest growth rate and largest average size, while the adjustable cyclic shape results in the most uniform particle size distribution, leading to the highest-quality diamond growth. Consequently, we chose to deposit high-quality polycrystalline diamond films using the adjustable cyclic-shaped substrate holder [35].

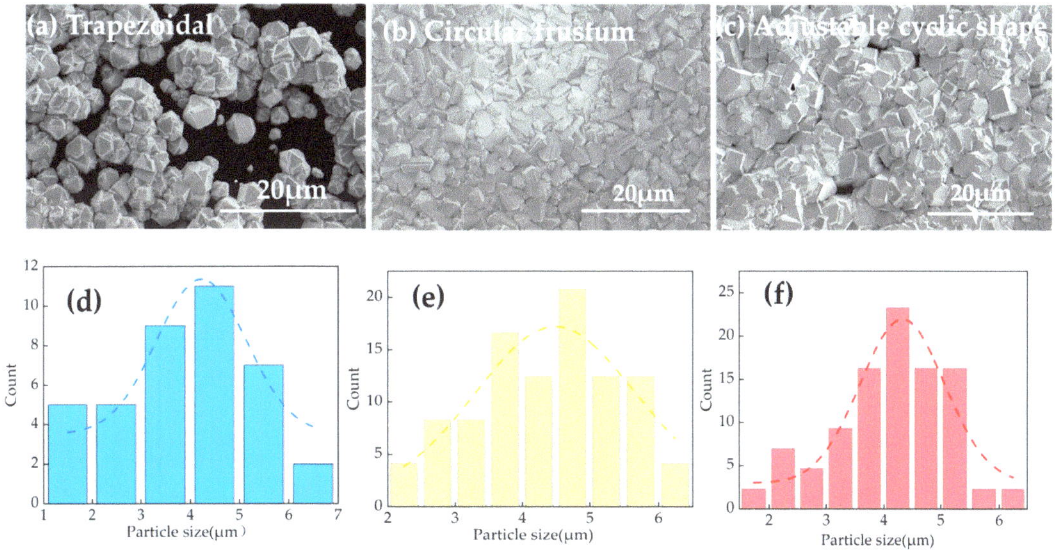

Figure 8. (**a**–**c**) The surface morphologies of polycrystalline diamond prepared on three different substrate supports and (**d**–**f**) the corresponding particle size distribution maps (The dashed line represents the Gaussian function fitted to obtain the normal distribution curve).

Temperature is also critical for polycrystalline diamond growth. We used infrared temperature measurements to monitor substrate growth through a viewing window at the top of the chamber. As shown in Figure 9, the center of the trapezoidal substrate holder has the highest temperature of 1136 K, while the lowest temperature at the edges is 935 K. This significant temperature difference leads to different diamond growth rates. Rapid changes in pressure and power can lead to silicon rupture. These results are consistent with the simulation results shown in Figure 5b, indicating that an inhomogeneous temperature field distribution is detrimental to the growth of polycrystalline diamond. In contrast, the temperature difference of the adjustable ring substrate holder is very small, which is consistent with the results in Figure 5d. Therefore, this study demonstrates the feasibility of the adjustable cyclic substrate holder and suggests a feasible route for growing large-scale high-quality polycrystalline diamond films.

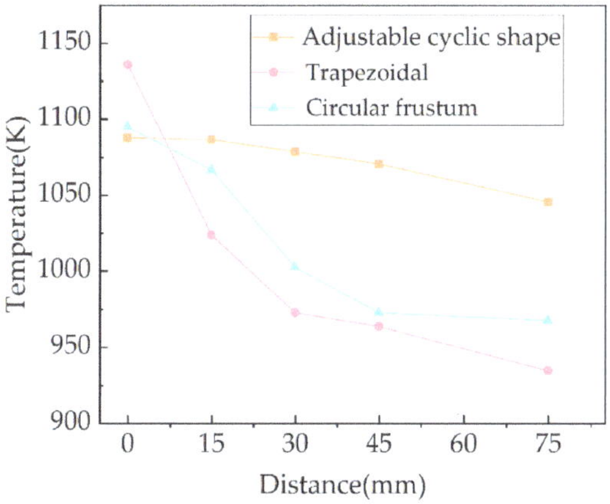

Figure 9. Temperature from the center of the substrate silicon wafer to the outside during growth of different substrate holders.

Additionally, to evaluate the quality of polycrystalline diamond produced by the three different substrate holders, Raman spectroscopy was conducted, as shown in Figure 10. The residual stress in the diamond films can be calculated based on the shift in the Raman peak using the following equation:

$$\sigma = m(\nu - \nu_0) \tag{11}$$

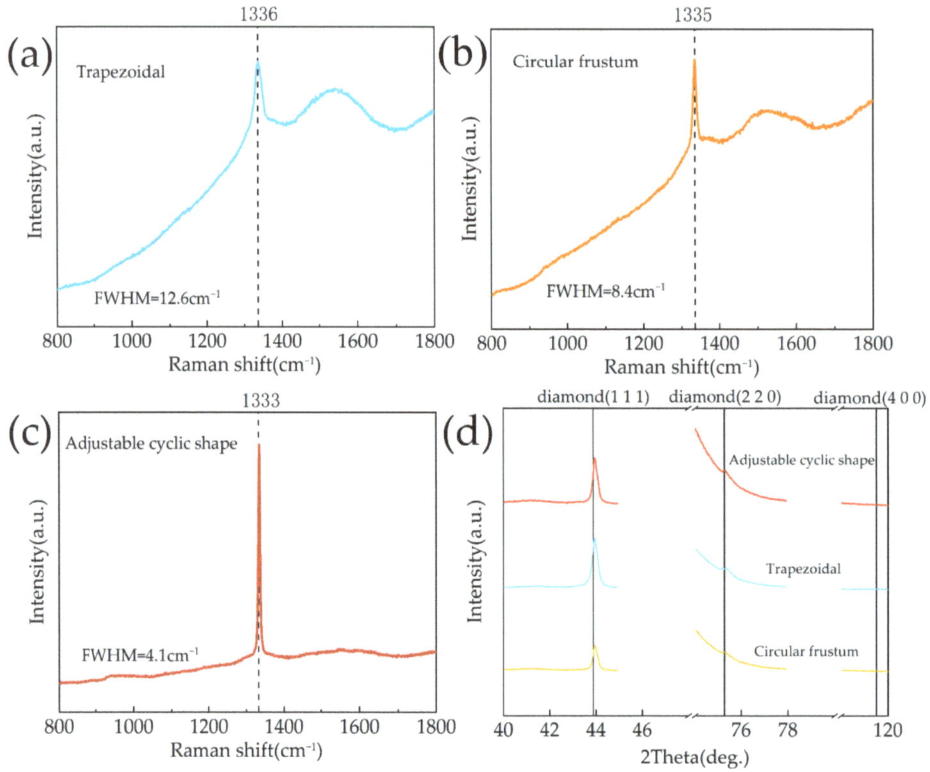

Figure 10. Raman spectra (**a**–**c**) and XRD diffraction patterns (**d**) of polycrystalline diamond prepared on different substrate supports.

In this equation, m is $-0.567\,\text{GPa}/\text{cm}^{-1}$, ν_0 is the wavenumber of natural, stress-free diamond (1332 cm^{-1}), and ν is the measured wavenumber of the diamond. Therefore, the residual stress in all samples predominantly manifests as compressive stress. Specifically, the trapezoidal sample exhibits a residual stress of 2.268 GPa, the circular frustum sample shows a stress of 1.701 GPa, and the adjustable cyclic-shaped sample presents a stress of 0.567 GPa. Additionally, the FWHM of the characteristic Raman peak, which is a critical parameter for assessing diamond quality, is 12.6 cm^{-1}, 8.4 cm^{-1}, and 4.1 cm^{-1}, respectively, for these samples. In the XRD patterns, only the (111) and (220) diffraction peaks are observed, with the (400) peak displaying the highest intensity, as shown in Figure 10d. Notably, the adjustable cyclic-shaped substrate holder exhibits the lowest FWHM at 0.27°, indicating the highest crystal quality. This indicates that the shape of the substrate holder significantly impacts the quality of the diamond. Among the tested designs, the diamond produced using the adjustable cyclic-shaped substrate holder exhibited the highest quality.

Owing to the etching effect of hydrogen plasma on the graphite sp^2 phase, polycrystalline diamond can undergo high-quality growth [21]. In comparison, the trapezoidal substrate holder resulted in a larger graphite phase. The conical substrate holder also contained some graphite phase. This is due to the rapid growth of diamond, which leads to the deposition of graphite that is not completely etched away. Overall, using an adjustable circular substrate holder ensures a uniform distribution of the electric field, plasma electron density, and temperature, making it feasible to grow high-quality three-inch polycrystalline diamond. The experimental results are consistent with the simulation results, validating the reliability of the simulation method used in this study.

4. Conclusions

This study investigates the impact of three distinct substrate holder shapes—trapezoidal, circular frustum, and adjustable cyclic—on the quality of polycrystalline diamond films. Using a frequency-domain transient solver, simulations were conducted to model the electric field and plasma distribution for each holder shape. Changes in the holder shape resulted in corresponding variations in the electric field and plasma distribution within the chamber. The trapezoidal shape enhanced secondary plasma formation, but the plasma ball did not fully cover the surface. The circular frustum shape reduced secondary plasma, minimizing energy loss and achieving the highest growth rate. However, it also introduced a graphite phase, which diminished the quality of the polycrystalline diamond films. To overcome these challenges, an adjustable cyclic substrate holder was proposed. This design suppressed secondary plasma, and the gas temperature distribution above the substrate evolved from a sombrero shape to a more uniform profile, improving the uniformity of temperature and hydrogen species density. Infrared temperature measurements showed that the trapezoidal holder exhibited a significant temperature gradient from the center to the edge, which hindered uniform growth. In contrast, the adjustable cyclic shape demonstrated the smallest temperature variation across the substrate surface, leading to more uniform growth. The polycrystalline diamond films deposited using the adjustable cyclic holder exhibited the lowest residual stress (0.567 GPa) and the narrowest FWHM of 4.1 cm^{-1}, predominantly oriented along the (111) crystal plane. The experimental results validated the simulation findings and confirmed the effectiveness of the adjustable cyclic substrate holder. This design offers a viable approach to enhancing the performance of existing MPCVD systems, facilitating the production of high-quality, large-area polycrystalline diamond films.

Author Contributions: Conceptualization, S.W.; methodology, S.W., J.L. and K.G.; software, S.W., J.Z. and K.G.; formal analysis, S.W.; investigation, K.G.; data curation, L.L. and L.H.; writing—original draft, S.W. and K.G.; writing—review and editing, J.B., K.G., C.Z. and Q.W.; project administration, K.G., Q.W. and J.B.; funding acquisition, K.G. and Q.W. All authors have read and agreed to the published version of the manuscript.

Funding: This research was funded by the Guangdong Basic and Applied Basic Research Foundation, China (Grant No. 2023B1515120033), the National Natural Science Foundation of China (Grant Nos. 52102037, 52302037), the General Program of the Chongqing Municipal Science and Technology Commission (Grant No. CSTB2022NSCQ-MSX0495), the Chongqing Municipal Education Commission Science and Technology Research Project (Grant Nos. KJQN202000742, KJQN202100733), and the Research and Innovation Program for Graduate Students in Chongqing (Grant No. CYS240525).

Informed Consent Statement: Not applicable.

Data Availability Statement: The original contributions presented in this study are included in the article; further inquiries can be directed to the corresponding authors.

Acknowledgments: Grateful acknowledgment is extended to the Jihua Laboratory Testing Center for their assistance with the SEM images and Raman spectroscopy. Appreciation is also given to Jingming Zhu and Jiafeng Li for their valuable help in conducting numerical simulations and experiments.

Conflicts of Interest: The authors declare no conflicts of interest.

References

1. Sun, D.-R.; Wang, G.; Li, Y.; Yu, Y.; Shen, C.; Wang, Y.; Lu, Z. Laser drilling in silicon carbide and silicon carbide matrix composites. *Opt. Laser Technol.* **2024**, *170*, 110166. [CrossRef]
2. Castelletto, S.; Boretti, A. Gallium Nitride Nanomaterials and Color Centers for Quantum Technologies. *ACS Appl. Nano Mater.* **2024**, *7*, 5862–5877. [CrossRef]
3. Yang, H.; Sun, J.; Wang, H.; Li, H.; Yang, B. A review of oriented wurtzite-structure aluminum nitride films. *J. Alloys Compd.* **2024**, *989*, 174330. [CrossRef]
4. Chen, J.; Du, X.; Luo, Q.; Zhang, X.; Sun, P.; Zhou, L. A Review of Switching Oscillations of Wide Bandgap Semiconductor Devices. *IEEE Trans. Power Electron.* **2020**, *35*, 13182–13199. [CrossRef]
5. Mazumder, S.K.; Voss, L.F.; Dowling, K.M.; Conway, A.; Hall, D.; Kaplar, R.J.; Pickrell, G.W.; Flicker, J.; Binder, A.T.; Chowdhury, S.; et al. Overview of Wide/Ultrawide Bandgap Power Semiconductor Devices for Distributed Energy Resources. *IEEE J. Emerg. Sel. Top. Power Electron.* **2023**, *11*, 3957–3982. [CrossRef]
6. Zhang, C.; Vispute, R.D.; Fu, K.; Ni, C. A review of thermal properties of CVD diamond films. *J. Mater. Sci.* **2023**, *58*, 3485–3507. [CrossRef]
7. Jia, X.; Wei, J.; Huang, Y.; Shao, S.; An, K.; Kong, Y.; Liu, J.; Chen, L.; Li, C. Fabrication of low stress GaN-on-diamond structure via dual-sided diamond film deposition. *J. Mater. Sci.* **2021**, *56*, 6903–6911. [CrossRef]
8. Jaeger, W. CVD Diamond Films—Synthesis, Microstructure, Applications. *Microsc. Microanal.* **2017**, *23* (Suppl. S1), 2260–2261. [CrossRef]
9. Siyi, C.; Juping, T.; Ke, H.; Siwu, S.; Zhiliang, Y.; Peng, L.; Jinlong, L.; Liangxian, C.; Junjun, W.; Kang, A.; et al. Uniform Growth of Two-inch MPCVD Optical Grade Diamond Film. *J. Inorg. Mater.* **2023**, *38*, 1413–1419.
10. Wang, Q.; Wu, G.; Liu, S.; Gan, Z.; Yang, B.; Pan, J. Simulation-Based Development of a New Cylindrical-Cavity Microwave-Plasma Reactor for Diamond-Film Synthesis. *Crystals* **2019**, *9*, 320. [CrossRef]
11. Yu, L.; Zhang, Y.Q.; Jin, X.R.; Lee, Y.P. Dual-band unidirectional reflectionless phenomenon based on high-order plasmon resonators in plasmonic waveguide system. *J. Nonlinear Opt. Phys. Mater.* **2022**, *31*, 2350002. [CrossRef]
12. Hong, S.P.; Lee, K.-i.; You, H.J.; Jang, S.O.; Choi, Y.S. Scanning Deposition Method for Large-Area Diamond Film Synthesis Using Multiple Microwave Plasma Sources. *Nanomaterials* **2022**, *12*, 1959. [CrossRef] [PubMed]
13. Mukherjee, D.; Oliveira, F.; Trippe, S.C.; Rotter, S.; Neto, M.; Silva, R.; Mallik, A.K.; Haenen, K.; Zetterling, C.-M.; Mendes, J.C. Deposition of diamond films on single crystalline silicon carbide substrates. *Diam. Relat. Mater.* **2020**, *101*, 107625. [CrossRef]
14. Bolshakov, A.P.; Ralchenko, V.G.; Yurov, V.Y.; Shu, G.; Bushuev, E.V.; Khomich, A.A.; Ashkinazi, E.E.; Sovyk, D.N.; Antonova, I.A.; Savin, S.S.; et al. Enhanced deposition rate of polycrystalline CVD diamond at high microwave power densities. *Diam. Relat. Mater.* **2019**, *97*, 107466. [CrossRef]
15. Wang, H.; Wang, C.; Wang, X.; Sun, F. Effects of carbon concentration and gas pressure with hydrogen-rich gas chemistry on synthesis and characterizations of HFCVD diamond films on WC-Co substrates. *Surf. Coat. Technol.* **2021**, *409*, 126839. [CrossRef]
16. Yudin, I.B.; Emelyanov, A.A.; Plotnikov, M.Y.; Rebrov, A.K.; Timoshenko, N.I. Influence of nitrogen on the synthesis of diamonds during gas-jet HWCVD deposition. *Fuller. Nanotub. Carbon Nanostructures* **2022**, *30*, 126–132. [CrossRef]
17. Wang, H.; Shen, X.; Wang, X.; Sun, F. Simulation and experimental researches on the substrate temperature distribution of the large-capacity HFCVD setup for mass-production of diamond coated milling tools. *Diam. Relat. Mater.* **2020**, *101*, 107610. [CrossRef]
18. Yang, Z.; Liu, Y.; Guo, Z.; Wei, J.; Liu, J.; Chen, L.; Li, C. Deposition of uniform diamond films on three dimensional Si spheres by using faraday cage in MPCVD reactor. *Diam. Relat. Mater.* **2024**, *142*, 110767. [CrossRef]
19. Sedov, V.; Martyanov, A.; Altakhov, A.; Popovich, A.; Shevchenko, M.; Savin, S.; Zavedeev, E.; Zanaveskin, M.; Sinogeykin, A.; Ralchenko, V.; et al. Effect of Substrate Holder Design on Stress and Uniformity of Large-Area Polycrystalline Diamond Films Grown by Microwave Plasma-Assisted CVD. *Coatings* **2020**, *10*, 939. [CrossRef]
20. Curto, S.; Taj-Eldin, M.; Fairchild, D.; Prakash, P. Microwave ablation at 915 MHz vs 2.45 GHz: A theoretical and experimental investigation. *Med. Phys.* **2015**, *42*, 6152–6161. [CrossRef]
21. Yang, D.; Guo, L.; Wang, B.; Jin, S.; Zhu, J.; Zhai, M. Hydrogen plasma characteristics in a microwave chemical vapor deposition chamber. *Mater. Sci. Eng. B* **2023**, *292*, 116422. [CrossRef]
22. Liu, S.; Jiang, L.; Li, H.; Zhao, J.; Wang, K.; Wang, H.; Li, T.; Zhou, Y.; Ghannouchi, F.M.; Hu, B. Design of a coaxial and compact TM01–TE01 mode converter based on helical corrugated waveguide for high-power microwave system. *AIP Adv.* **2022**, *12*, 115120. [CrossRef]
23. Xiao, R.; Chen, K.; Wang, H.; Wang, D.; Shi, Y.; Gao, L. Theoretical calculation and particle-in-cell simulation of a multi-mode relativistic backward wave oscillator operating at low magnetic field. *Phys. Plasmas* **2022**, *29*, 043103. [CrossRef]
24. Weng, J.; Xiong, L.W.; Wang, J.H.; Dai, S.Y.; Man, W.D.; Liu, F. Investigation of depositing large area uniform diamond films in multi-mode MPCVD chamber. *Diam. Relat. Mater.* **2012**, *30*, 15–19. [CrossRef]
25. An, K.; Chen, L.; Liu, J.; Zhao, Y.; Yan, X.; Hua, C.; Guo, J.; Wei, J.; Hei, L.; Li, C.; et al. The effect of substrate holder size on the electric field and discharge plasma on diamond-film formation at high deposition rates during MPCVD. *Plasma Sci. Technol.* **2017**, *19*, 095505. [CrossRef]

26. Zhao, W.; Teng, Y.; Tang, K.; Zhu, S.; Yang, K.; Duan, J.; Huang, Y.; Chen, Z.; Ye, J.; Gu, S. Significant suppression of residual nitrogen incorporation in diamond film with a novel susceptor geometry employed in MPCVD. *Chin. Phys. B* **2022**, *31*, 118102. [CrossRef]
27. Hassouni, K.; Grotjohn, T.A.; Gicquel, A. Self-consistent microwave field and plasma discharge simulations for a moderate pressure hydrogen discharge reactor. *J. Appl. Phys.* **1999**, *86*, 134–151. [CrossRef]
28. Li, X.-J.; Zhou, S.; Chen, G.; Wang, D.-S.; Pei, N.; Guo, H.-L.; Nie, F.-M.; Zhang, X.; Feng, S. Systematic research on the performance of self-designed microwave plasma reactor for CVD high quality diamond. *Def. Technol.* **2018**, *14*, 373–379. [CrossRef]
29. Zhang, Y.; Yu, S.; Gao, J.; Ma, Y.; He, Z.; Hei, H.; Zheng, K. Design and simulation of a novel MPCVD reactor with three-cylinder cavity. *Vacuum* **2022**, *200*, 111055. [CrossRef]
30. Weng, J.; Liu, F.; Wang, Z.T.; Guo, N.F.; Fan, F.Y.; Yang, Z.; Wang, J.B.; Wang, H.; Xiong, L.W.; Zhao, H.Y.; et al. Investigation on the preparation of large area diamond films with 150–200 mm in diameter using 915 MHz MPCVD system. *Vacuum* **2023**, *217*, 112543. [CrossRef]
31. Harris, S.J.; Goodwin, D.G. Growth on the reconstructed diamond (100) surface. *J. Phys. Chem.* **1993**, *97*, 23–28. [CrossRef]
32. Silva, F.; Hassouni, K.; Bonnin, X.; Gicquel, A. Microwave engineering of plasma-assisted CVD reactors for diamond deposition. *J. Phys. Condens. Matter* **2009**, *21*, 364202. [CrossRef] [PubMed]
33. Hassouni, K.; Silva, F.; Gicquel, A. Modelling of diamond deposition microwave cavity generated plasmas. *J. Phys. D Appl. Phys.* **2010**, *43*, 153001. [CrossRef]
34. Yang, B.; Shen, S.; Zhang, L.; Shen, Q.; Zhang, R.; Zhang, Y.; Gan, Z.; Liu, S. Study on diamond temperature stability during long-duration growth via MPCVD under the influence of thermal contact resistance. *J. Appl. Crystallogr.* **2022**, *55*, 240–246. [CrossRef]
35. Chen, K.; Tao, T.; Hu, W.; Ye, Y.; Zheng, K.; Ye, J.; Zhi, T.; Wang, X.; Liu, B.; Zhang, R. High-speed growth of high-quality polycrystalline diamond films by MPCVD. *Carbon Lett.* **2023**, *33*, 2003–2010. [CrossRef]

crystals

MDPI

Article

Plasma-Enhanced Atomic Layer Deposition of Hematite for Photoelectrochemical Water Splitting Applications

Thom R. Harris-Lee [1,2], Andrew Brookes [1], Jie Zhang [2], Cameron L. Bentley [2], Frank Marken [1] and Andrew L. Johnson [1,*]

[1] Department of Chemistry, University of Bath, Claverton Down, Bath BA2 7AY, UK; trhl21@bath.ac.uk (T.R.H.-L.); ab2926@bath.ac.uk (A.B.); f.marken@bath.ac.uk (F.M.)

[2] School of Chemistry, Monash University, Clayton, VIC 3800, Australia; jie.zhang@monash.edu (J.Z.); cameron.bentley@monash.edu (C.L.B.)

[*] Correspondence: a.l.johnson@bath.ac.uk

Abstract: Hematite (α-Fe_2O_3) is one of the most promising and widely used semiconductors for application in photoelectrochemical (PEC) water splitting, owing to its moderate bandgap in the visible spectrum and earth abundance. However, α-Fe_2O_3 is limited by short hole-diffusion lengths. Ultrathin α-Fe_2O_3 films are often used to limit the distance required for hole transport, therefore mitigating the impact of this property. The development of highly controllable and scalable ultrathin film deposition techniques is therefore crucial to the application of α-Fe_2O_3. Here, a plasma-enhanced atomic layer deposition (PEALD) process for the deposition of homogenous, conformal, and thickness-controlled α-Fe_2O_3 thin films (<100 nm) is developed. A readily available iron precursor, dimethyl(aminomethyl)ferrocene, was used in tandem with an O_2 plasma co-reactant at relatively low reactor temperatures, ranging from 200 to 300 °C. Optimisation of deposition protocols was performed using the thin film growth per cycle and the duration of each cycle as optimisation metrics. Linear growth rates (constant growth per cycle) were measured for the optimised protocol, even at high cycle counts (up to 1200), confirming that all deposition is 'true' atomic layer deposition (ALD). Photoelectrochemical water splitting performance was measured under solar simulated irradiation for pristine α-Fe_2O_3 deposited onto FTO, and with a α-Fe_2O_3-coated TiO_2 nanorod photoanode.

Keywords: hematite; atomic layer deposition; plasma; ferrocene

Citation: Harris-Lee, T.R.; Brookes, A.; Zhang, J.; Bentley, C.L.; Marken, F.; Johnson, A.L. Plasma-Enhanced Atomic Layer Deposition of Hematite for Photoelectrochemical Water Splitting Applications. *Crystals* **2024**, *14*, 723. https://doi.org/10.3390/cryst14080723

Academic Editors: Miłosz Grodzicki, Damian Wojcieszak and Michał Mazur

Received: 3 July 2024
Revised: 5 August 2024
Accepted: 10 August 2024
Published: 13 August 2024

1. Introduction

Atomic layer deposition (ALD) is a highly precise and controlled gas phase method for the deposition of thin films onto a range of substrates, first developed independently within groups in the Soviet Union (1960s) and in Finland (1974) [1–3]. It has since been applied across a range of semiconductor-based technology that requires ultrathin, controlled, and conformal coatings [4]. One such use is in photoelectrochemical (PEC) devices, which can employ ALD to fabricate heterojunctions [5], protective layers [6], and blocking layers [7]. PEC devices are typically highly nanostructured with complex morphologies, making ALD perfectly suited for the deposition of conformal layers onto these high-aspect-ratio arrangements without losing or filling-in the desired architecture.

The conventional method of ALD, also known as thermal ALD, uses elevated substrate temperatures and simple oxidants (e.g., H_2O, H_2O_2, O_2 and O_3) to drive surface reactions in each half-cycle [1,8]. However, these common oxidants are often not suitable co-reactants for efficient deposition on less reactive precursor compounds, significantly restricting the viable precursor options. An alternative method that is more effective on less reactive precursors is plasma-enhanced ALD (PEALD) [9,10], which utilises a strongly oxidising plasma (commonly O_2 or N_2 plasmas) as the co-reactant in the second half-cycle, providing access to lower substrate/deposition temperatures, faster reactions, shorter purge times, and a shorter initial nucleation delay before linear growth rates [9]. However, limitations of PEALD make its use situational and make it require stringent optimisation and characterisation protocols, including lower thin film uniformity, plasma-induced damage to the growing film or underlayers, and undesired surface reactions and defect formation [11–13].

ALD of α-Fe_2O_3 has been given significantly less attention compared to the alternative common metal oxide materials due to its low growth rates, low precursor volatilities and reactivities, and narrow temperature windows for ALD growth [14–18]. A range of Fe_2O_3 precursors have been studied and reported, including $Fe(thd)_3$ [19], $Fe_2(O^tBu)_6$ [20], bis(2,4-methylpentadienyl)iron [21], $FeCl_3$ [22], $Fe(acac)_3$ [23], and Fe(btmsa) [24], using either H_2O, H_2O_2, O_2, or O_3 as a co-reactant, and sharing similar optimal deposition conditions [19,25–28].

While precursors such as $FeCl_3$ are hampered by corrosion issues in the ALD chamber [22], precursors such as ferrocene [29] have received attention because of the combined features of its low cost, high level of availability, high air moisture, and thermal stability, making it an ideal candidate for scale-up [19,27]. However, even Ferrocene-based precursors require long deposition durations due to poor reactivity, often using O_3 oxidant as the most reactive co-reactant commonly used in thermal ALD. It therefore makes sense to instead use highly reactive oxygen plasma as the oxidant (i.e., PEALD). Despite this, to the best of the authors' knowledge, only two studies have been published using PEALD to deposit α-Fe_2O_3. Detavernier et al. [30] used a tertiary butyl ferrocene precursor with O_2 plasma to produce crystalline, pure α-Fe_2O_3 films within the temperature range 250–400 °C. Jeong et al. [31] compared thermal ALD and PEALD for the deposition of α-Fe_2O_3 from bis(*N*,*N*′-dibutylacetamidinato)iron(II) precursor, revealing that PEALD-grown α-Fe_2O_3 possessed lower surface roughness and better crystallinity than the equivalent film grown by thermal ALD. Given the advantages of PEALD for overcoming limitations within α-Fe_2O_3 ALD, it is surprising that no additional literature exists on similar α-Fe_2O_3 PEALD processes.

Hematite (α-Fe_2O_3) possesses high stability under a large pH range, non-toxicity, high natural abundance, low cost, and narrow bandgap of 2.0–2.2 eV, allowing absorption of most of the visible light spectrum (up to 620 nm) [32]. This combination of properties has made it one of the most promising and best-researched materials for use in PEC water splitting, however, its application in energy storage devices such as batteries [33–36] and supercapacitors [37–39], sensors that detect the presence of certain gases or chemicals [40–45], and PEC solar cells has also been explored [46–48]. The predominant limitation of α-Fe_2O_3 is its short hole-diffusion lengths, restricting its use to ultra-thin films or highly nanostructured morphologies such as nanowires to minimise diffusion distances [49–51]. A TiO_2 underlayer is known to improve the PEC performance of α-Fe_2O_3 on FTO substrate as α-Fe_2O_3 and FTO have a significant lattice mismatch, resulting in a poor interface for electron transport [50]. In fact, a study by Grätzel et al. used an O_3 co-reactant in a thermal ALD process to conformally coat TiO_2 nanorods with a α-Fe_2O_3 overlayer, which significantly enhanced the PEC performance of the electrode relative to both films individually [52]. Here we have developed for the first time a viable PEALD process for the fabrication of hematite thin films using a widely commercially available precursor system. This study represents our initial research in this area.

2. Results

Dimethyl(aminomethyl)ferrocene (DMAMFc) has been previously reported by Grätzel et al. [52] as a precursor in an ALD process using ozone (O_3) oxidant as the co-reactant. It was therefore deemed a promising option for application within a PEALD regime. Due to the differing technique and instruments, it was important to optimise the complete process for PEALD, and not simply the second, plasma half-cycle parameters.

Thermogravimetric analysis (TGA) was performed on DMAMFc to reveal its volatilisation behaviour and suitability for ALD (Figure 1). A single and relatively fast mass loss event was measured, showing clean volatilisation (reaching ~0% wt%) initiating at 108.2 °C (temperature after 1% mass loss) and completing at 217.7 °C. The absence of any signs of decomposition reveals the high thermal stability of the DMAMFc precursor, a vital property for ALD precursors to ensure no decomposition (CVD) processes occur during precursor pulse half-cycles. Indeed, possession of both suitable volatility and high thermal stability indicates the promise of DMAMFc precursor towards ALD.

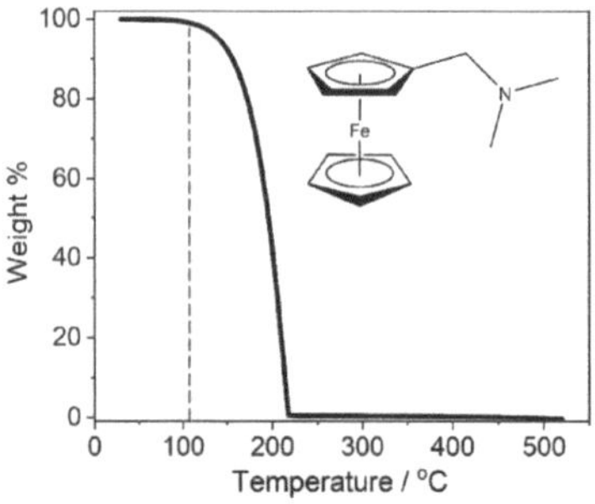

Figure 1. Thermogravimetric analysis of DMAMFc, measured under an Ar flow between 30 and 520 °C at a constant ramp rate of 5 °C min^{-1}, with volatilisation onset marked with a red dashed line.

2.1. PEALD Process Development

All optimisation figures herein have used a 'standard deposition procedure' developed throughout the work as the most optimised parameters. Unless otherwise stated, all figures will vary only according to the discussed parameters from this procedure. The standard deposition procedure was: 90 °C pot temperature, 240 °C chamber temperature, {0.5/3.0/5.0 s} × 3 vapour boost/precursor-pulse/purge sequence in the precursor pulse half-cycle, 5 s plasma pulse (100 W), and 10 s plasma purge. Preliminary depositions were ineffective and yielded negligible Fe_2O_3 thicknesses due to the slow rate of precursor uptake. In order to effect precursor transfer, a vapour boost protocol was introduced, specifically, the pressure within the precursor pot was increased by a short (<1 s) nitrogen pulse immediately before exposure to the vacuum, resulting in a more forceful initial response and increased agitation of the liquid precursor, and therefore the vapour formation and uptake.

Growth per cycle (GPC) values (obtained by measurement of the final thickness of a grown film with a known number of cycles) from no boost to 1 s boost sequences (Figure 2a) reveal the significant enhancement in performance provided, and that the DMAMFc PEALD process is likely primarily limited by the precursor extraction from the pot into the chamber.

The GPC begins to plateau after 0.5 s, an indication that precursor uptake has passed the threshold where it is the limiting factor for deposition, and the process is now limited by the reaction rate at the sample surface. A duration of 0.5 s was therefore chosen to maximise uptake and minimise cycle duration.

Precursor pulse lengths from 0.5 to 5 s, all with a 0.5 s boost, were trialled, yielding an apparent GPC saturation regime from 2 s onward, a classic indication of self-limiting ALD growth (blue trace and inset, Figure 2b). However, the GPC value at the plateau (~0.29 Å) was significantly lower than expected for an ALD process with complete surface saturation in each cycle. Considering boosts were essential for extraction of the precursor, multiple pulses, each with its own boost, were trialled within the same precursor half-cycle, effectively refreshing the pot to the boost pressure after each individual pulse (black trace, Figure 2b). Using the multi-pulse half-cycles, the GPC increased significantly for the same total precursor pulse lengths and continued a relatively linear increase as total pulse length increased further, even reaching total pulse lengths of 30 s (10 consecutive 0.5 s boosts + 3 s pulse) without a plateau.

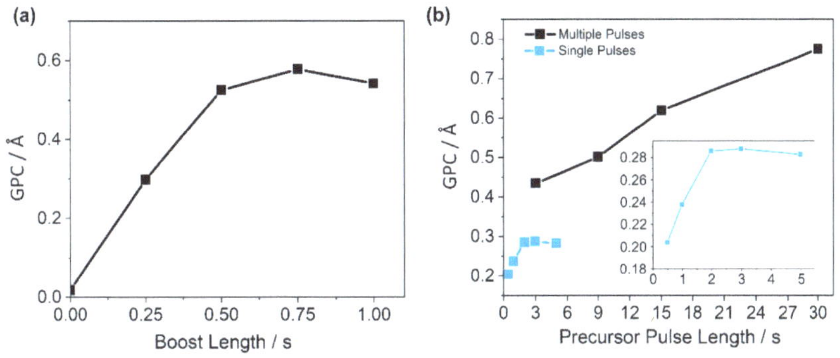

Figure 2. Optimisation of DMAMFc PEALD pulse sequence process, comparing measured growth rate to (**a**) DMAMFc delivery boost pulse, and (**b**) DMAMFc delivery pulse for a single boost pulse before plasma step (blue, zoomed on inset), and multiple boost pulse sequences before the plasma step (black). All depositions completed with the defined 'standard protocol' except for the single varied parameter.

These data clearly show that the precursor is unusually difficult to volatilise, reaching a point after initial exposure to the vacuum where no, or significantly limited, volatilisation occurs, despite the continuation of the vacuum environment. While it is conventional to increase pulse duration until saturated ALD growth is seen, 30 s total pulse lengths consumed large quantities of precursor and required long deposition periods of 110 s per cycle. It was therefore concluded that a higher pulse duration was impractical for employment in both a research and commercial environment.

A linear increase in GPC with increasing pot temperature was seen (Figure 3a), suggesting that the rate of monolayer deposition within the DMAMFc deposition was limited by the amount of precursor pulsed into the chamber. Temperatures above 120 °C were not tested as a previous report using this precursor identified a reduction in photoactivity of the resulting film with pot temperatures at 120 °C and higher [52]. The temperature of the reaction chamber was optimised to reside within the self-limiting growth ALD window. After trialling six points between 200 and 300 °C, the GPC plateaus at 240 and 260 °C before decreasing at 280 and 300 °C, likely due to desorption from the sample surface (Figure 3b).

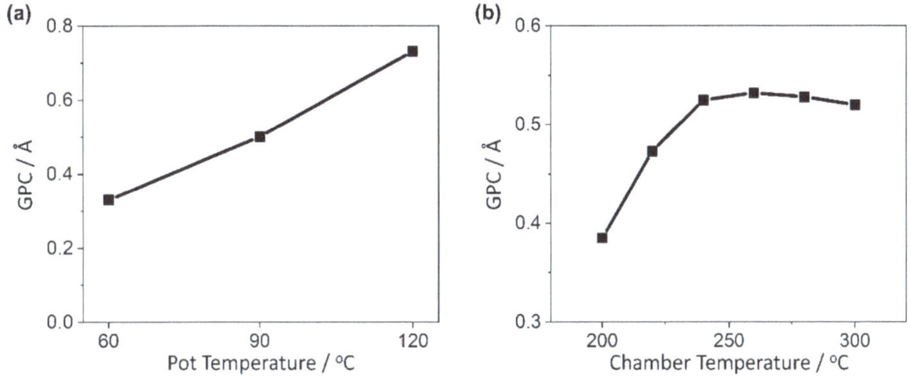

Figure 3. Optimisation of DMAMFc PEALD temperatures, comparing measured growth rate with (**a**) pot temperature, and (**b**) chamber temperature. All depositions completed with the defined 'standard protocol' except for the single varied parameter.

There are no signs of CVD contribution, therefore the precursor can be used at all these temperatures. The optimal temperature was chosen to be 240 °C, despite its 0.007 Å cycle^{-1}

lower GPC than 260 °C (0.525 compared to 0.532 Å cycle^{-1}, respectively). Lower temperature processes are preferred due to the reduced energy requirement and greater flexibility in choice of substrate and co-reactant, hence the marginal GPC difference does not justify the increased temperature.

Oxygen plasma was produced using an RF generator, with power ranging from 0 W (resulting in only an O_2 gas pulse) to 200 W (Figure 4a). For 0 W, no O_2 plasma will be produced, and the negligible GPC (0.0065 Å) revealed that O_2 does not act as an oxidant to functionalise the surface during co-reactant half-cycles, confirming that all deposits within this study have been solely PEALD, and not a combination of PEALD and thermal ALD. The GPC shows a positive correlation with plasma power from 50 to 150 W, as increased power generated more O_2 plasma and therefore more of the surface can be functionalised. However, instead of plateauing as plasma power increased further (tending towards complete surface functionalisation), the GPC decreased significantly at 200 W. It is likely that the growing Fe_2O_3 layers were sensitive to excessive O_2 plasma exposure, resulting in damage or removal of the surface upon the plasma pulse half cycle. In support of this, a similar trend was identified for the plasma pulse duration parameter (Figure 4b). A GPC of zero with no plasma pulse confirmed that no CVD occurred, followed by a positive correlation of increasing GPC with plasma pulse duration up to a 7 s pulse period, after which a detrimental effect was, again, observed. This is not a new observation, with similar having been reported previously for a range of different materials [53,54].

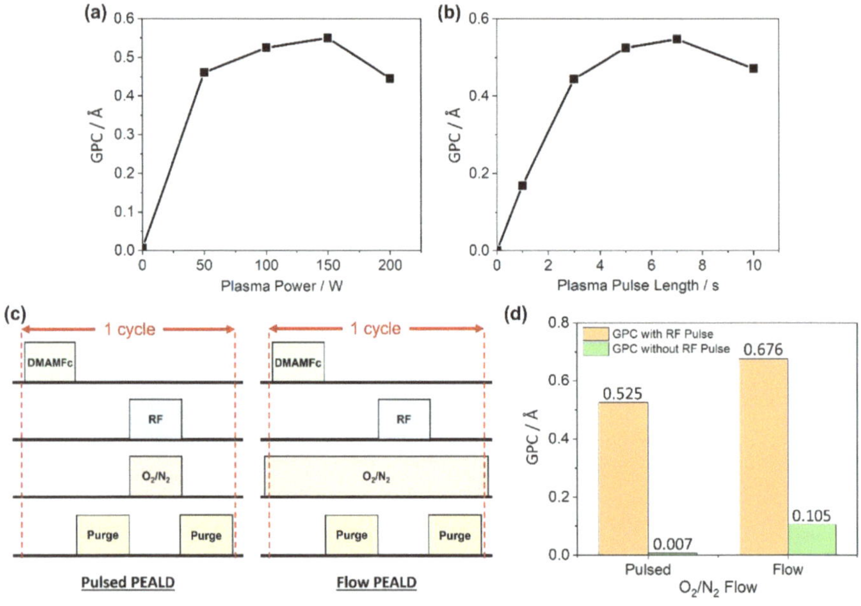

Figure 4. Optimisation of DMAMFc PEALD plasma parameters, comparing measured growth rate with (**a**) plasma power, and (**b**) plasma pulse length. All depositions completed with the defined 'standard protocol' except for the single varied parameter. (**c**) Comparison of pulse/purge/pulse/purge protocols for pulsed vs. flow plasma gas input, and (**d**) measured growth from pulsed vs. flow plasma gas protocols, with and without RF pulse.

All deposition processes used thus far have been in a pulsed plasma gas regime, meaning that the plasma gas in the co-reactant half-cycle is pulsed into the chamber simultaneously with the RF power on/off (Figure 4c). Alternatively, the plasma gas can be flowed continuously into the chamber throughout the entire ALD cycle window, and only the RF generator will be pulsed in the co-reactant half-cycle (i.e., O_2 gas flow is constant, but O_2 plasma is only generated during this second half-cycle). When the Fe_2O_3 GPC from each process is compared, it appears that the flow regime is more effective (Figure 4d).

However, when the same regimes were trialled without the RF pulse (i.e., expecting zero deposition as no O_2 plasma was generated), the flow regime still had a significant GPC value (0.105 Å cycle^{-1}). This indicates that CVD processes were occurring, likely due to the reaction between O_2 and the DMAMFc precursor at the elevated temperature during precursor pulsing. The flow regime was therefore deemed unsuitable for this process.

The GPC of a series of ALD processes with varying pulse sequences, pot temperatures, and chamber temperatures were recorded (Table 1, Figure 5). For practical application, a high GPC is not useful if each cycle is excessive in duration. A new efficiency parameter (η) was therefore introduced to account for this, calculated by dividing the *GPC* by the duration of each full ALD cycle, as in Equation (1):

$$\eta = GPC\ ()/t_{cycle}\ (s) \tag{1}$$

where η is the deposition efficiency and t_{cycle} is the duration of one full ALD cycle. Processes 1–4 were all single-pulse sequences, and all possessed lower efficiencies than all 3× (and even 5×) multi-pulse sequences, once again confirming the effectiveness of employing multiple boost/pulse sequences within the same half-cycle, despite the time it adds to the cycle. While 5× and 10× boost/pulse sequences yield the highest GPC values (processes 8 and 10, respectively), the increase is not enough to mitigate the resulting extended cycle durations, evidenced by the lower η values. A half-cycle of 3× boost/pulse sequences is therefore considered the optimal parameter out of this dataset. Process 9 is clearly optimal for deposition, showcasing the greatest GPC and efficiency; however, as previously discussed, a precursor pot temperature of 120 °C is known to decrease the photoactivity of the resulting Fe_2O_3 film [52]. Consequently, processes 6 and 7 emerge as the optimal conditions. Process 7 consumed more precursor while only exhibiting a marginal increase in GPC and η, thus process 6 was concluded to be the most optimal among parameters tested.

Table 1. PEALD processes with varying pulse sequences and temperatures, all using 100 W RF generator power, a 5 s plasma gas and RF pulse, and a 10 s purge.

Process Number	Sequence/s (Boost/Pulse) × *n*	Temp/°C (Pot/Chamber)	GPC/Å cycle^{-1}	η/10^{-3} Å s^{-1}
1	(0.5/1.0) × 1	60/240	0.13	4.16
2	(2.0/5.0) × 1	60/240	0.24	6.46
3	(0.5/3.0) × 1	90/240	0.29	8.60
4	(0.5/3.0) × 1	90/300	0.29	8.72
5	(0.5/1.0) × 3	90/240	0.43	9.75
6	(0.5/3.0) × 3	90/240	0.53	10.40
7	(1.0/3.0) × 3	90/240	0.54	10.42
8	(0.5/3.0) × 5	90/240	0.62	9.17
9	(0.5/3.0) × 3	120/240	0.73	14.49
10	(0.5/3.0) × 10	120/240	0.78	7.05

The linearity of growth rate seen in both total film thickness and GPC trends (Figure 6), using the optimised parameters from process 6, reveal the high degree of control this process has for the deposition of Fe_2O_3. An approximately decreasing GPC with increasing cycles suggests it may become more difficult to grow the film as thickness increases, however, the drop is negligible for the number of cycles being investigated. These trends are indicative of a pure ALD process, without any CVD contribution. The process is therefore an effective method for depositing ultrathin (<100 nm) Fe_2O_3 films, and can now be applied to PEC systems to study its photoactivity and performance, both alone and in a simple heterojunction.

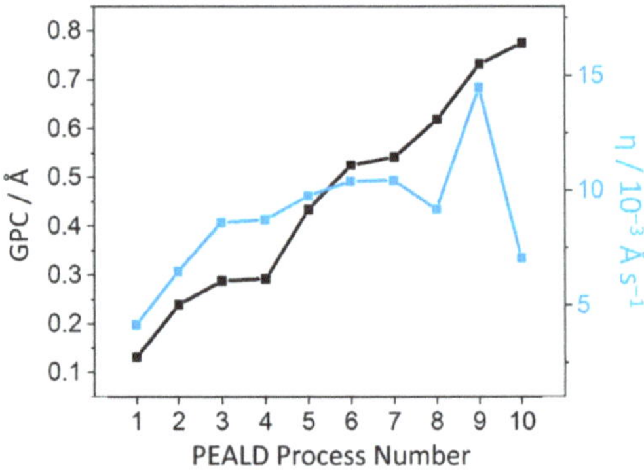

Figure 5. Growth rates and corresponding deposition efficiencies (η) for a series of PEALD processes.

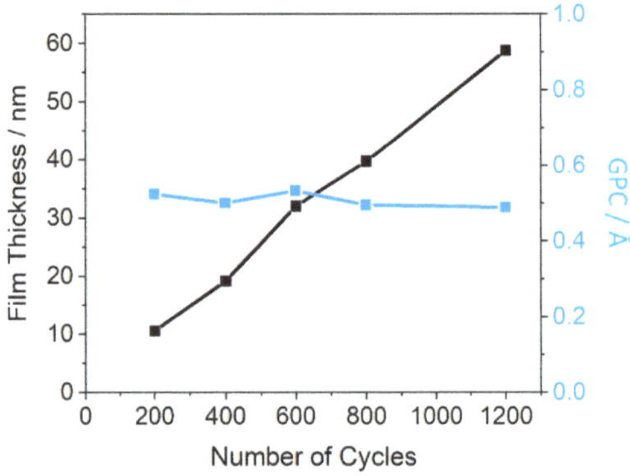

Figure 6. Growth rate and total film thickness measurements for films deposited by PEALD using process 6 across a range of total cycle numbers.

2.2. Thin Film Physical Characterisation

All initial deposition processes were initially performed using silica wafers as a thin film substrate. However, for PEC application, deposition onto a transparent conducting oxide (TCO) substrate was required. FTO-coated (TEC-15) glass slides were therefore used as the substrate. Selected samples were annealed at 500 °C for 4 h after deposition, which resulted in a change of colour from dark brown to orange/red (Figure 7), indicating formation of crystalline α-Fe_2O_3 only after annealing. An annealing temperature of 500 °C was used in other studies and was chosen here as it is below the glass transition temperature of TEC-15 (564 °C) [55].

Figure 7. Appearance (from left to right) of: initial FTO-coated glass substrate, as-deposited 60 nm thin films, annealed 60 nm thin film (note that the silver area at the top is silver paint and not part of the sample).

Grazing incidence XRD (GIXRD) was employed on an annealed Fe_2O_3 thin film (500 °C for 4 h) and a pristine FTO-coated glass substrate control with an incidence angle (ω) of 5°. Spectra were obtained within the 20–90° (Figure 8a) and the 30–40° (Figure 8b) 2θ ranges—the latter also performed on an as-deposited (unannealed) Fe_2O_3 thin film. In the resulting spectra, a clear and unambiguous peak at 35.6° (matching the position of the expected (110) α-Fe_2O_3 peak) was present for the annealed Fe_2O_3 thin film, but was absent for both the FTO and, interestingly, unannealed Fe_2O_3. As-deposited Fe_2O_3 must therefore be amorphous, only becoming crystalline after annealing at 500 °C. Amorphous materials can possess improved conductivity in bulk, compared to their crystalline counterparts, and would be particularly effective as an outermost layer (which is complementary to the use of ALD as the conformal coating tool) in an electrode due to the higher surface energy of amorphous materials, therefore increasing the electrocatalytic activities. However, the study of the amorphous thin film is beyond the scope of this preliminary work, and therefore all samples herein have been annealed at 500 °C for 4 h and will therefore be denoted as α-Fe_2O_3.

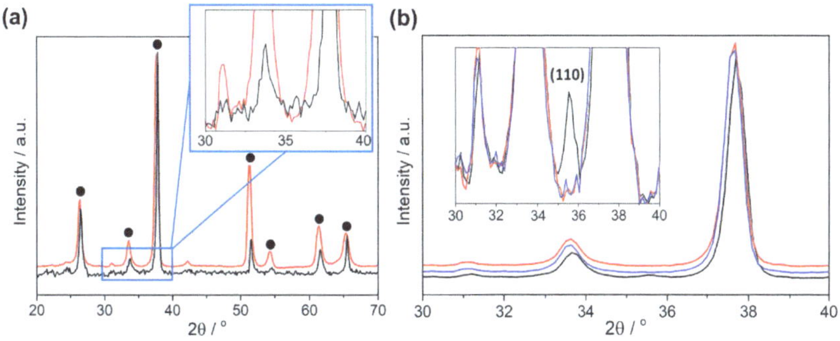

Figure 8. Grazing incidence X-ray diffraction patterns for pristine FTO (red), (blue) as-deposited Fe_2O_3 on FTO, and annealed Fe_2O_3 on FTO (black), measured at an X-ray incidence angle of 5° for ~12 h within a 2θ range of (**a**) 20–90° and (**b**) 30–40°. Insets show zoomed views of the peak of interest with XRD patterns overlayed. Peaks associated with FTO are marked with a black dot.

SEM cross-sectional images coupled with EDX were used to identify a 60 nm α-Fe_2O_3 layer on a pristine FTO-coated glass substrate (Figure 9). It is difficult to confirm the thickness of the α-Fe_2O_3 due to the poor resolution at such high magnifications, and the lack of distinction between the conformal α-Fe_2O_3 coating and the FTO layer, however, it was approximated to be 75 nm. It could therefore be concluded that the α-Fe_2O_3 growth is more favored on FTO than silica (i.e., there is more surface coverage as the surface is closer to saturation in each individual precursor pulse half-cycle). EDX data defined a clear, flat, and uniform Fe layer on top of the Sn (FTO) layer.

To demonstrate the conformal deposition and layering potential of the PEALD process, a 30 nm- and 60 nm-α-Fe_2O_3 thin film was deposited onto a previously studied [5,56,57] photoanode consisting of TiO_2 nanorods grown onto FTO substrate (herein denoted as TiO_2/Fe_2O_3-30 nm and TiO_2/Fe_2O_3-60 nm, respectively). Similar TiO_2 nanorods have previously shown enhanced PEC performance after α-Fe_2O_3 ALD coating [52].

Top-down SEM images of the TiO_2 (Figure 10a) and TiO_2/Fe_2O_3-60 nm (Figure 10b) show a seemingly successful conformal coating onto the nanorod morphologies. The roughness of the tips of the TiO_2 nanorods vanishes after the α-Fe_2O_3 coating, resulting in a far smoother morphology. The diameter of the nanorods does not appear to change significantly after coating (~70–100 nm), suggesting that the growth rate measured on the silica substrate may be greater compared to a TiO_2 surface, although the nature of smoothing rough edges and measuring on nanorods makes it difficult to accurately measure and/or confirm this. For clarity, both heterojunction samples will continue to be referred to as TiO_2/Fe_2O_3-30 nm and TiO_2/Fe_2O_3-60 nm herein.

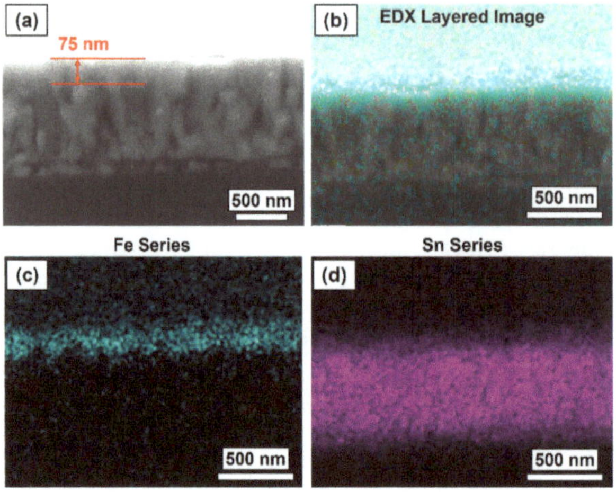

Figure 9. (**a**) FE-SEM cross-section image of 60 nm-α-Fe_2O_3 deposited onto FTO-coated glass substrate (approximate α-Fe_2O_3 layer indicated in red). EDX data for 60 nm-α-Fe_2O_3 deposited onto FTO-coated glass substrate: (**b**) electron image overlayed with Fe series from EDX analysis (teal), (**c**) Fe map across electron image, and (**d**) Sn map across the electron image.

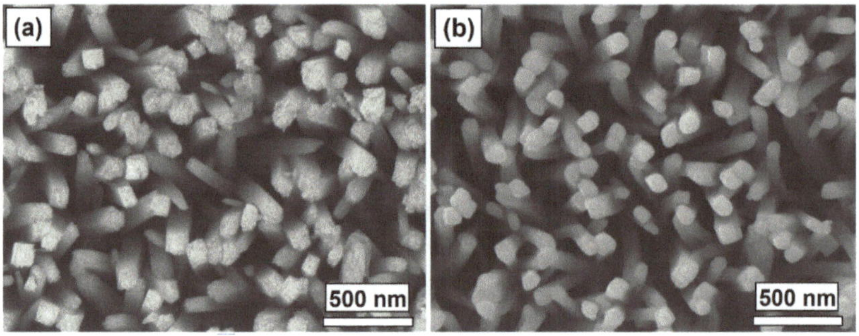

Figure 10. FE-SEM surface images of (**a**) TiO_2-nanorods, and (**b**) TiO_2/Fe_2O_3-60nm, deposited onto FTO-coated glass substrate.

The highly directional nature of plasma often results in a significant decrease in the conformality of a PEALD coating compared to thermal ALD [9]. Given the high aspect-ratio of the TiO_2 nanorod structure, it was necessary to study the depth of the Fe_2O_3 coating on the nanorods via EDX of thin film cross-sections (Figure 11). Qualitative analysis of Figure 11a,c reveals a clear decrease in the abundance of Fe further down the length of the nanorod structures. Interestingly, the thicker Fe_2O_3 coating (TiO_2/Fe_2O_3-60 nm, Figure 11c) shows greater relative Fe abundance near the lower region of the nanostructures, revealing that Fe_2O_3 must be deposited across the entire structure, but without complete coverage further down the rod length. This suggests that longer precursor pulse half cycles could provide greater conformality by ensuring complete monolayer coverage of the lower nanorod regions. However, as discussed previously, it is impractical to increase the deposition durations further.

UV/Vis spectroscopy revealed a sharp absorption onset at 415 nm for the as-deposited TiO_2 film (Figure 12a), corresponding to a 3.01 eV bandgap evaluated using the corresponding Tauc plot (Figure 12b), and the values matched those of the expected rutile polymorph. Interestingly, α-Fe_2O_3 showed no characteristic absorbance onset, however absorbance does appear to flatten at >600 nm, which matches the expected bandgap. It is likely that the film was too thin for clear absorption onset peaks, particularly when considering the low absorption coefficient of α-Fe_2O_3 (requiring a 375 nm thickness to absorb 95% of 550 nm incident light) [50]. The absorption spectra from the TiO_2/Fe_2O_3 heterojunction contained two distinct absorption peaks at 595 nm and 405 nm, representing bandgaps of 3.08 and 2.09 eV, respectively. The 2.09 eV peak is within the expected range for α-Fe_2O_3 (1.9–2.2 eV) [58], and the 3.08 eV peak is between values for anatase (3.2 eV) and rutile (3.0 eV) TiO_2. The presence of two TiO_2 phases is unexpected, as it is known from a previous study [56] that the TiO_2 nanorods are purely rutile when deposited onto FTO.

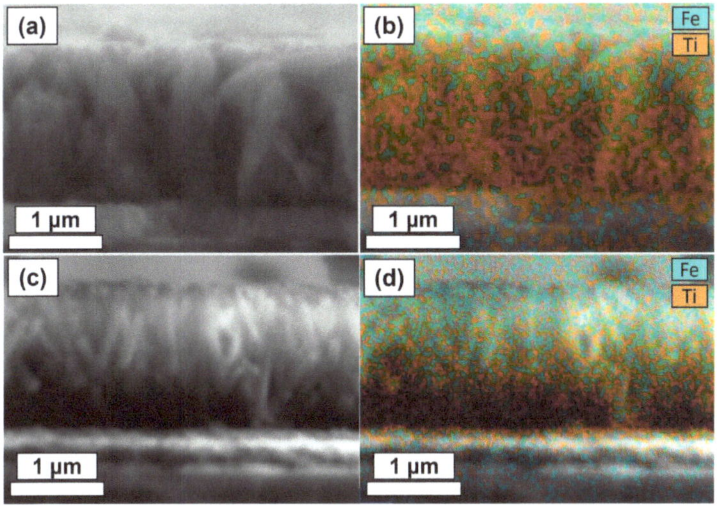

Figure 11. (**a**) FE-SEM cross-section image of TiO_2/Fe_2O_3-30 nm, (**b**) electron image of TiO_2/Fe_2O_3-30 nm overlayed with Fe (blue) and Ti (orange) series from EDX analysis, (**c**) FE-SEM cross-section image of TiO_2/Fe_2O_3-60 nm, and (**d**) electron image of TiO_2/Fe_2O_3-60 nm overlayed with Fe (blue) and Ti (orange) series from EDX analysis.

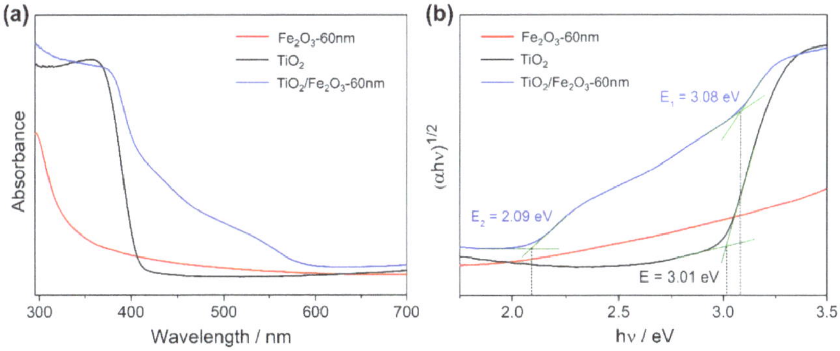

Figure 12. (**a**) Optical absorption spectra for: (black) TiO_2, (red) α-Fe_2O_3, (blue) TiO_2/Fe_2O_3. (**b**) Tauc plots for each sample with extrapolated indirect bandgaps as annotated. All thin films were deposited onto FTO-coated glass.

Standard reflection-XRD of TiO_2 and TiO_2/Fe_2O_3-60 nm was performed to further study the TiO_2 phases present, and the possibility of rutile conversion to anatase during the ALD process. XRD patterns (Figure 13) contained only rutile peaks for TiO_2 environments, which contrasts the bandgap measured via Tauc analysis (rutile bandgap expected: ~3.00 eV, measured: 3.08 eV). It is therefore likely that the slight widening is a result of either: (i) long-term (24 h) exposure to the vacuum environment at 240 °C with regular oxygen plasma pulses has caused a widening of the bandgap [56], or (ii) the uncertainty associated with the Tauc analysis is ≥0.08 eV. Potential a-Fe_2O_3 peaks are visible in the TiO_2/Fe_2O_3 pattern (red highlight), however, the similar sizes relative to the background noise, and the overlap of 2θ positions with rutile and FTO peaks, limit the certainty of identification.

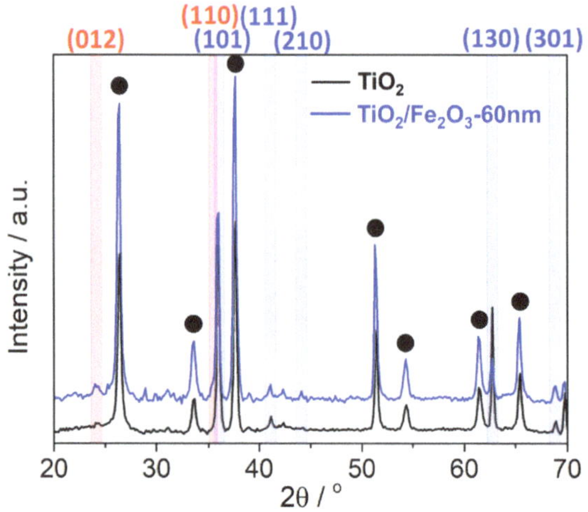

Figure 13. X-ray diffraction pattern for the pristine TiO_2 nanorod and TiO_2/Fe_2O_3-60 nm samples, both grown onto FTO-coated glass substrate. Peak locations corresponding to rutile TiO_2 and a-Fe_2O_3 are highlighted in blue and red, respectively, and FTO is marked with a black dot.

2.3. Thin Film Photoelectrochemical Characterisation

PEC analysis on a pristine TiO_2 nanorod thin film, 30 nm α-Fe_2O_3, and the TiO_2/Fe_2O_3 heterojunction with 30 nm (Figure 14a,b) and 60 nm (Figure 14c,d) α-Fe_2O_3 thicknesses was performed under both front and rear illumination. A summary of the photocurrent densities for each photoanode at 1.23 V vs. RHE is provided in Table 2. Like many metal oxide electrodes, the pristine TiO_2 nanorod electrode showed greater performance under rear illumination than front—an indication that charge carrier transport within the material may be relatively slow with respect to the timescale for recombination.

Photocurrents for α-Fe_2O_3 were negligible from both front and rear illumination, likely due to the ultrathin layer resulting in no appreciable photon absorption, particularly relevant for α-Fe_2O_3 which possesses a relatively small absorption coefficient [50]. As expected, given this and the possible detrimental band alignment, the TiO_2/Fe_2O_3-30 nm thin film showed poorer photocurrents than pristine TiO_2, most noticeably for front-side (α-Fe_2O_3 first) illumination. In fact, the relative difference between front- and rear-side performance for TiO_2/Fe_2O_3-30 nm (2.8 vs. 20 μA cm^{-2}) was far greater than for TiO_2 (24 vs. 40 μA cm^{-2}) (Figure 14a,b). Interestingly, the TiO_2/Fe_2O_3-60 nm heterojunction had a front-side performance worse than TiO_2 (13 vs. 24 μA cm^{-2}), albeit still significantly higher than that for TiO_2/Fe_2O_3-30 nm, while the rear-side photocurrent density was over double (85 vs. 40 μA cm^{-2}) (Figure 14c,d). The addition of the α-Fe_2O_3 layer is therefore detrimental under front illumination for both thicknesses but can improve performance when illuminated from the rear for thicker (60 nm) coatings.

A cathodic baseline current is observed within all photocurrent density scans (Figure 14), with varying onsets for each sample, indicating that a reduction reaction becomes thermodynamically favourable at potentials more negative than the onset potential. The baseline shift is reproduceable with repeat experiments, and there is no oxidation reaction seen at high anodic potentials, hence it cannot be attributed to experimental error such as electrolyte contact with the silver paint or copper tape used to connect the electrode to the potentiostat.

It is therefore concluded that the reduction reaction must be either self-reduction, which will impact the stability of the thin film(s), or reduction of a species within the solution, for example, oxygen. The onset potential (vs. RHE) of the cathodic baseline for all samples is in the order: TiO_2 (0.29 V) < TiO_2/Fe_2O_3 (0.69 V) < Fe_2O_3-30 nm (0.9V), and the thickness of the Fe_2O_3 layer in the heterojunction does not impact the onset.

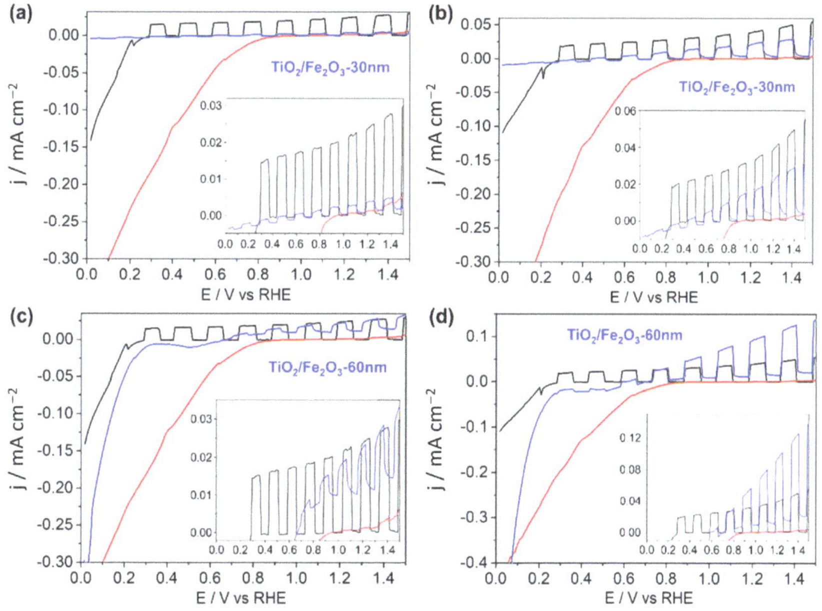

Figure 14. Linear sweep voltammograms under one sun-chopped AM 1.5 (**a**,**c**) front, (**b**,**d**) rear illumination for (black) TiO_2, (red) 30 nm α-Fe_2O_3, ((**a**,**b**); blue) TiO_2/Fe_2O_3-30nm, ((**c**,**d**); blue TiO_2/Fe_2O_3-60 nm). Insets contain zoomed-in views ignoring the cathodic current density baseline shift. All measurements performed in 1 M KOH (pH 13.7) with a 15 mV s^{-1} scan rate.

Table 2. Photocurrent densities at 1.23 V vs. RHE for photoanodes measured in Figure 14.

			$j_{photo}/\mu A\ cm^{-2}$	
Illumination Side	**TiO_2**	**30 nm α-Fe_2O_3**	**TiO_2/Fe_2O_3-30 nm**	**TiO_2/Fe_2O_3-60 nm**
Front	24	0.46	2.8	13
Rear	40	0.43	20	85

IPCE measurements for TiO_2 and TiO_2/Fe_2O_3-60 nm (the highest performing heterojunction) thin films were carried out to further investigate the photoelectronic impact of the α-Fe_2O_3 coating (Figure 15). TiO_2 possessed the expected trend given its large bandgap, the IPCE only significantly increasing above zero at <420 nm (2.95 eV) incident photon wavelength and peaking at 360 nm (3.44 eV). Over this range, the IPCE of the TiO_2/Fe_2O_3-60 nm-layered film was enhanced significantly, increasing from 8.5% to 43.5% at 360 nm. While this could be partly due to increased photon absorption, the α-Fe_2O_3 layer was significantly thinner than TiO_2, so considering that the wavelength was within the bandgap of TiO_2, the addition of α-Fe_2O_3 was not expected to significantly increase the total 360 nm photon absorption of the electrode. The increased performance must therefore be linked to improved charge separation due to the heterojunction formation, and therefore the band alignment must, in fact, become preferential (type-II heterojunction) for electron-hole separation upon semiconductor contact [59,60]. The narrower bandgap of α-Fe_2O_3 (measured as ~2.09 eV, Figure 12) results in IPCE above zero for incident wavelengths greater than the 420 nm cut-off seen in TiO_2. Interestingly, despite measuring the ~2.09 eV (593 nm) bandgap, the IPCE of TiO_2/Fe_2O_3-60nm remained above zero up to ~620 nm (2.0 eV), the lower limit of the accepted α-Fe_2O_3 bandgap range.

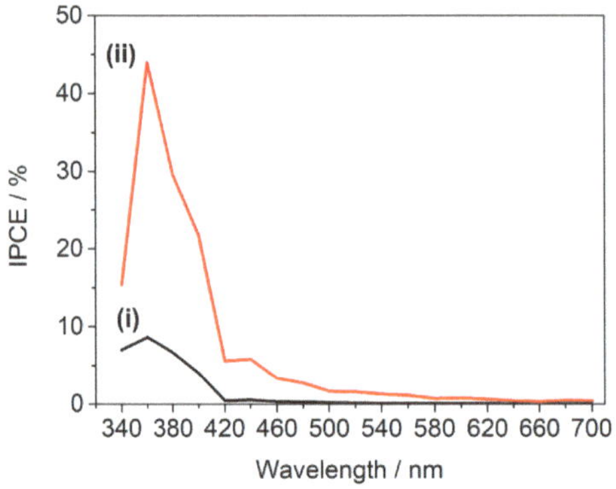

Figure 15. Incident photon-electron conversion efficiencies measured at 1.23 V vs. RHE under one sun AM 1.5 rear illumination for (i) as-deposited TiO_2, and (ii) TiO_2/Fe_2O_3-60 nm. All measurements carried out 1 M KOH (pH 13.7).

3. Conclusions

An effective deposition protocol for conformal, thickness-controlled α-Fe_2O_3 thin films was developed using PEALD, for which there are only two other reported processes. Pure ALD growth was confirmed by a constant GPC and linearly increasing film thickness with an increasing number of growth cycles, and an optimised deposition protocol was determined from several possible processes by comparison with the respective GPC and time efficiency values. Due to the poor uptake of the precursor, a vapour boost protocol was introduced and repeated within a single precursor pulse half-cycle to ensure sufficient GPC for film growth on a practical timescale, although, even with this addition, the deposition periods were high, which prevented film-thickness growth beyond 60 nm. Poor uptake is a common property of ferrocene-based precursors, hence it is of interest to expand and adapt this deposition protocol to a bespoke α-Fe_2O_3 precursor, tailor-made for this PEALD process (i.e., possessing greater volatility than commercially available ferrocene-based structures, improving uptake and reducing required pot temperatures), which is particularly relevant for PEALD compared to thermal ALD due to the greater flexibility in molecular precursor design afforded by the high-reactivity oxygen plasma co-reactant.

To confirm the conformality of the thin film deposition, a coating of α-Fe_2O_3 was deposited onto a photoanode consisting of TiO_2 nanorods. The resulting film did not show complete uniform and conformal deposition down the entire length of the nanorod, with decreasing Fe abundance towards the lower region of the nanorod structure. Photoelectrochemical water splitting measurements were performed on annealed α-Fe_2O_3, as well as the α-Fe_2O_3-coated TiO_2 nanorods and as-deposited TiO_2 nanorods for performance comparisons, revealing that the thickest (60 nm) α-Fe_2O_3 coating was the only electrode to compete with the performance of as-deposited TiO_2, possessing ca. half the photocurrent density at 1.23 V vs. RHE from the front-side illumination, but ca. double that from the rear-side illumination. Despite focusing on the annealed films here, the as-deposited Fe_2O_3 was found to be amorphous and would be of interest for further study in PEC application as a heterojunction. Overall, this study represents our initial results from the development of thin films of hematite using a PEALD process. Future work in this area should focus on the development of Fe_2O_3 films at lower temperatures on softer substrates, and will be reported elsewhere.

4. Experimental

Precursor thermal characterisation was performed using thermogravimetric analysis (TGA, PerkinElmer TGA 4000), heating the compound of interest between 50 and 520 °C at a constant ramp rate of 5 °C min^{-1} under a 20 mL min^{-1} argon flow. For isothermal measurements, the precursor was heated to 90 °C, held for 8 min, then heated to 95 °C at a constant ramp rate of 5 °C min^{-1}, and repeated until the final temperature had been reached.

All depositions were performed on a Beneq TFS-200 reactor using a direct, capacitively coupled plasma configuration. Dimethylaminomethyl ferrocene (DMAMFc, Alfa Aesar, >98%) was used without further purification. DMAMFc was kept in a HS300 stainless steel container and heated to 60, 90, or 120 °C. To avoid condensation of the precursor, ultrapure nitrogen (N_2) was used as a carrier gas and purging gas. O_2 and N_2 were used in the plasma system and maintained at 50 sccm and 200 sccm, respectively, throughout the deposition. The process was trialled in both pulsed and flow plasma gas regimes. A 13.56 MHz RF power source (CESAR 133, Advanced Energy) and impedance matching network (Navio, Advanced Energy) system were used to generate O_2 plasma. The process was systematically examined to optimise the precursor pulse and purging times. Depositions were completed using applied plasma powers ranging from 0 to 200 W, and a chamber temperature range of 75–325 °C.

Growth of α-Fe_2O_3 was primarily investigated by depositing it onto 150 mm Si (100) wafers. Resulting film thicknesses were measured using spectroscopic ellipsometry (Stokes LSE-USB ellipsometer), and crystallinity was measured by grazing incidence X-ray diffraction (GIXRD, STOE STADI P Bragg-Brentano geometry, CuK incident X-rays) at incidence angles of 0.5–5°. Absorption spectra were collected on samples deposited onto ultrasonic and plasma-cleaned, fluorine-doped tin oxide (FTO, AGC type U TCO glass)-coated glass, using a Cary 5000 UV-Vis-NIR spectrophotometer equipped with an integrating sphere and a centre mount sample holder to account for scattering and reflection contribution. High resolution images of the thin films were obtained by field emission scanning electron microscopy (FESEM, JEOL JSM-7900F), and elemental analysis was performed by energy dispersive X-ray spectroscopy (EDX, Oxford Instruments AZtec 170 mm^2 Ultim Max).

The TiO_2 nanocrystal array was grown on FTO-coated glass substrate by a hydrothermal synthesis reaction previously reported by Zhang et al. [34]. Briefly, an FTO-coated glass substrate was ultrasonically cleaned for 30 min in a 1:1 solution of ethanol and acetone, and placed inside a 100 mL Teflon-lined stainless-steel autoclave with the conductive FTO-coated side facing upwards. A solution containing 0.7 g of titanium(IV) butoxide in a 50 mL mixture of 1:1 concentrated hydrochloric acid (37%) and de-ionised water was stirred for 30 min and pipetted onto the FTO-coated substrate. The autoclave was sealed and heated in an oven at 180 °C for 3 h.

Electrochemical data were recorded with an Autolab electrochemical workstation (PGSTAT100) connected to a three-electrode electrochemical cell containing a platinum wire counter electrode (CE), 3 M Ag/AgCl reference electrode (RE), and thin film sample working electrode (WE). Simulated sunlight was generated from a 300 W Xenon lamp (Microsolar300 Beijing Perfectlight Technology Co. Ltd., AM 1.5 G, 100 mW cm^{-2}) and employed to provide chopped light (5 s on, 5 s off) incidents on the WE to demonstrate photocurrents and dark background currents across a potential range scanned with a scan rate of 15 mV s^{-1}. All measurements were performed in 1 M KOH (pH 13.7) electrolyte.

Author Contributions: A.L.J. conceived and designed the study and made a significant contribution to the editing and manuscript preparation alongside T.R.H.-L. T.R.H.-L. and A.B. performed the experimental work, collected data, and contributed to the data analysis and interpretation. T.R.H.-L. wrote and prepared an initial draft of the manuscript and contributed to the editing. A.L.J., F.M., C.L.B., and J.Z. contributed to the supervision of T.R.H.-L. and A.B., as well as to the interpretation of analysis of data, funding acquisition, visualization of the project, and the project's administration. All authors have read and agreed to the published version of the manuscript.

Funding: C.L.B. is the recipient of an Australian Research Council (ARC) Discovery Early Career Researcher Award (DECRA, project number: DE200101076), funded by the Australian Government.

Data Availability Statement: The data that support the findings of this study are available from the corresponding authors due to privacy.

Acknowledgments: This work has been supported by the University of Bath (UoBath) and Monash University (MonashU), both of which are thanked for the provision of a joint Bath–Monash Global PhD studentship to TRHL. The authors acknowledge the use of the instruments, and the scientific and technical assistance, at the Monash Centre for Electron Microscopy (MCEM), a Node of Microscopy Australia.

Conflicts of Interest: The authors declare no conflicts of interest.

References

1. George, S.M. Atomic Layer Deposition: An Overview. *Chem. Rev.* **2009**, *110*, 111–131. [CrossRef]
2. Malygin, A.A.; Drozd, V.E.; Malkov, A.A.; Smirnov, V.M.; From, V.B. Aleskovskii's "Framework" Hypothesis to the Method of Molecular Layering/Atomic Layer Deposition. *Chem. Vap. Depos.* **2015**, *21*, 216–240. [CrossRef]
3. Mallick, B.C.; Hsieh, C.-T.; Yin, K.-M.; Gandomi, Y.A.; Huang, K.-T. Review—On Atomic Layer Deposition: Current Progress and Future Challenges. *ECS J. Solid State Sci. Technol.* **2019**, *8*, N55–N78. [CrossRef]
4. Puurunen, R.L. Surface chemistry of atomic layer deposition: A case study for the trimethylaluminum/water process. *J. Appl. Phys.* **2005**, *97*, 121301. [CrossRef]
5. Innocent, J.W.F.; Napari, M.; Johnson, A.L.; Harris-Lee, T.R.; Regue, M.; Sajavaara, T.; MacManus-Driscoll, J.L.; Marken, F.; Alkhalil, F. Atomic scale surface modification of TiO_2 3D nano-arrays: Plasma enhanced atomic layer deposition of NiO for photocatalysis. *Mater. Adv.* **2021**, *2*, 273–279. [CrossRef]
6. O'Donnell, S.; Jose, F.; Shiel, K.; Snelgrove, M.; McFeely, C.; McGill, E.; O'Connor, R. Thermal and plasma enhanced atomic layer deposition of ultrathin TiO_2 on silicon from amide and alkoxide precursors: Growth chemistry and photoelectrochemical performance. *J. Phys. D Appl. Phys.* **2022**, *55*, 085105. [CrossRef]
7. Hu, H.; Dong, B.; Hu, H.; Chen, F.; Kong, M.; Zhang, Q.; Luo, T.; Zhao, L.; Guo, Z.; Li, J.; et al. Atomic Layer Deposition of TiO_2 for a High-Efficiency Hole-Blocking Layer in Hole-Conductor-Free Perovskite Solar Cells Processed in Ambient Air. *ACS Appl. Mater. Interfaces* **2016**, *8*, 17999–18007. [CrossRef] [PubMed]
8. Ponraj, J.S.; Attolini, G.; Bosi, M. Review on Atomic Layer Deposition and Applications of Oxide Thin Films. *Crit. Rev. Solid State Mater. Sci.* **2013**, *38*, 203–233. [CrossRef]
9. Profijt, H.B.; Potts, S.E.; van de Sanden, M.C.M.; Kessels, W.M.M. Plasma-Assisted Atomic Layer Deposition: Basics, Opportunities, and Challenges. *J. Vac. Sci. Technol. A Vac. Surf. Film.* **2011**, *29*, 050801. [CrossRef]
10. Kim, H.; Oh, I.-K. Review of plasma-enhanced atomic layer deposition: Technical enabler of nanoscale device fabrication. *Jpn. J. Appl. Phys.* **2014**, *53*, 03DA01. [CrossRef]
11. Takagi, T. Ion–surface interactions during thin film deposition. *J. Vac. Sci. Technol. A Vac. Surf. Film.* **1984**, *2*, 382–388. [CrossRef]
12. Ren, H.; Nishi, Y.; Shohet, J.L. Changes to Charge and Defects in Dielectrics from Ion and Photon Fluences during Plasma Exposure. *Electrochem. Solid-State Lett.* **2011**, *14*, H107. [CrossRef]
13. Oviroh, P.O.; Akbarzadeh, R.; Pan, D.; Coetzee, R.A.M.; Jen, T.-C. New development of atomic layer deposition: Processes, methods and applications. *Sci. Technol. Adv. Mater.* **2019**, *20*, 465–496. [CrossRef] [PubMed]
14. Lie, M.; Fjellvåg, H.; Kjekshus, A. Growth of Fe_2O_3 thin films by atomic layer deposition. *Thin Solid Film.* **2005**, *488*, 74–81. [CrossRef]
15. Rooth, M.R.; Johansson, A.; Kukli, K.; Aarik, J.; Boman, M.; Hårsta, A. Atomic Layer Deposition of Iron Oxide Thin Films and Nanotubes using Ferrocene and Oxygen as Precursors. *Chem. Vap. Depos.* **2008**, *14*, 67–70. [CrossRef]
16. Lin, Y.; Xu, Y.; Mayer, M.T.; Simpson, Z.I.; McMahon, G.; Zhou, S.; Wang, D. Growth of p-Type Hematite by Atomic Layer Deposition and Its Utilization for Improved Solar Water Splitting. *J. Am. Chem. Soc.* **2012**, *134*, 5508–5511. [CrossRef] [PubMed]
17. Riha, S.C.; Klahr, B.M.; Tyo, E.C.; Seifert, S.; Vajda, S.; Pellin, M.J.; Hamann, T.W.; Martinson, A.B.F. Atomic Layer Deposition of a Submonolayer Catalyst for the Enhanced Photoelectrochemical Performance of Water Oxidation with Hematite. *ACS Nano* **2013**, *7*, 2396–2405. [CrossRef] [PubMed]
18. Emery, J.D.; Schlepütz, C.M.; Guo, P.; Riha, S.C.; Chang, R.P.H.; Martinson, A.B.F. Atomic Layer Deposition of Epitaxial Iron Oxides for Photoelectrochemical Water Oxidation. *ECS Meet. Abstr.* **2015**, *MA2015-02*, 1719. [CrossRef]
19. Li, X.; Fan, N.C.; Fan, H.J. A Micro-pulse Process of Atomic Layer Deposition of Iron Oxide Using Ferrocene and Ozone Precursors and Ti-Doping. *Chem. Vap. Depos.* **2013**, *19*, 104–110. [CrossRef]
20. Bachmann, J.; Jing; Knez, M.; Barth, S.; Shen, H.; Mathur, S.; Gösele, U.; Nielsch, K. Ordered Iron Oxide Nanotube Arrays of Controlled Geometry and Tunable Magnetism by Atomic Layer Deposition. *J. Am. Chem. Soc.* **2007**, *129*, 9554–9555. [CrossRef]
21. Riha, S.C.; Racowski, J.M.; Lanci, M.P.; Klug, J.A.; Hock, A.S.; Martinson, A.B.F. Phase Discrimination through Oxidant Selection in Low-Temperature Atomic Layer Deposition of Crystalline Iron Oxides. *Langmuir* **2013**, *29*, 3439–3445. [CrossRef] [PubMed]

22. Klug, J.A.; Becker, N.G.; Riha, S.C.; Martinson, A.B.F.; Elam, J.W.; Pellin, M.J.; Proslier, T. Low temperature atomic layer deposition of highly photoactive hematite using iron(iii) chloride and water. *J. Mater. Chem. A* **2013**, *1*, 11607–11613. [CrossRef]
23. de Ridder, M.; van de Ven, P.C.; van Welzenis, R.G.; Brongersma, H.H.; Helfensteyn, S.; Creemers, C.; Van Der Voort, P.; Baltes, M.; Mathieu, M.; Vansant, E.F. Growth of Iron Oxide on Yttria-Stabilized Zirconia by Atomic Layer Deposition. *J. Phys. Chem. B* **2002**, *106*, 13146–13153. [CrossRef]
24. Selvaraj, S.; Moon, H.; Yun, J.-Y.; Kim, D.-H. Iron oxide grown by low-temperature atomic layer deposition. *Korean J. Chem. Eng.* **2016**, *33*, 3516–3522. [CrossRef]
25. Klahr, B.; Gimenez, S.; Fabregat-Santiago, F.; Hamann, T.; Bisquert, J. Water Oxidation at Hematite Photoelectrodes: The Role of Surface States. *J. Am. Chem. Soc.* **2012**, *134*, 4294–4302. [CrossRef] [PubMed]
26. Klahr, B.M.; Martinson, A.B.F.; Hamann, T.W. Photoelectrochemical Investigation of Ultrathin Film Iron Oxide Solar Cells Prepared by Atomic Layer Deposition. *Langmuir* **2011**, *27*, 461–468. [CrossRef] [PubMed]
27. Martinson, A.B.F.; DeVries, M.J.; Libera, J.A.; Christensen, S.T.; Hupp, J.T.; Pellin, M.J.; Elam, J.W. Atomic Layer Deposition of Fe2O3 Using Ferrocene and Ozone. *J. Phys. Chem. C* **2011**, *115*, 4333–4339. [CrossRef]
28. Van Bui, H.; Grillo, F.; van Ommen, J.R. Atomic and molecular layer deposition: Off the beaten track. *Chem. Commun.* **2017**, *53*, 45–71. [CrossRef]
29. Astruc, D. Why is Ferrocene so Exceptional? *Eur. J. Inorg. Chem.* **2016**, *2017*, 6–29. [CrossRef]
30. Ramachandran, R.K.; Dendooven, J.; Detavernier, C. Plasma enhanced atomic layer deposition of Fe_2O_3 thin films. *J. Mater. Chem. A* **2014**, *2*, 10662–10667. [CrossRef]
31. Choi, B.; Park, G.-W.; Jeong, J.-R.; Jeon, N. Comparative Study of Thermal and Plasma-Enhanced Atomic Layer Deposition of Iron Oxide Using Bis(N,N′-di-butylacetamidinato)iron(II). *Nanomaterials* **2023**, *13*, 1858. [CrossRef] [PubMed]
32. Chen, Q.; Fan, G.; Fu, H.; Li, Z.; Zou, Z. Tandem photoelectrochemical cells for solar water splitting. *Adv. Phys. X* **2018**, *3*, 1487267. [CrossRef]
33. Wang, L.; Liang, K.; Wang, G.; Yang, Y. Interface-engineered hematite nanocones as binder-free electrodes for high-performance lithium-ion batteries. *J. Mater. Chem. A* **2018**, *6*, 13968–13974. [CrossRef]
34. Yang, Y. A mini-review: Emerging all-solid-state energy storage electrode materials for flexible devices. *Nanoscale* **2020**, *12*, 3560–3573. [CrossRef] [PubMed]
35. Wang, K.; Cao, J.; Gao, J.; Zhao, J.; Jiang, W.; Ahmad, W.; Jiang, J.; Ling, M.; Liang, C.; Chen, J. Unveiling the structure-activity relationship of hollow spindle-like α-Fe_2O_3 nanoparticles via phosphorus doping engineering for enhanced lithium storage. *Sustain. Mater. Technol.* **2023**, *38*, e00744. [CrossRef]
36. Yu, L.; Zhou, X.; Lu, L.; Wu, X.; Wang, F. Recent Developments of Nanomaterials and Nanostructures for High-Rate Lithium Ion Batteries. *ChemSusChem* **2020**, *13*, 5361–5407. [CrossRef]
37. Lu, X.; Zeng, Y.; Yu, M.; Zhai, T.; Liang, C.; Xie, S.; Balogun, M.S.; Tong, Y. Oxygen-Deficient Hematite Nanorods as High-Performance and Novel Negative Electrodes for Flexible Asymmetric Supercapacitors. *Adv. Mater.* **2020**, *32*, 3148–3155. [CrossRef]
38. Lai, F.; Feng, J.; Heil, T.; Wang, G.-C.; Adler, P.; Antonietti, M.; Oschatz, M. Strong metal oxide-support interactions in carbon/hematite nanohybrids activate novel energy storage modes for ionic liquid-based supercapacitors. *Energy Storage Mater.* **2019**, *20*, 188–195. [CrossRef]
39. Yadav, A.A.; Deshmukh, T.B.; Deshmukh, R.V.; Patil, D.D.; Chavan, U.J. Electrochemical supercapacitive performance of Hematite α-Fe_2O_3 thin films prepared by spray pyrolysis from non-aqueous medium. *Thin Solid Film.* **2016**, *616*, 351–358. [CrossRef]
40. Hjiri, M.; Algessair, S.; Dhahri, R.; Mirzaei, A.; Neri, G. Gas sensing properties of hematite nanoparticles synthesized via different techniques. *RSC Adv.* **2024**, *14*, 17526–17534. [CrossRef]
41. Garcia, D.; Picasso, G.; Hidalgo, P.; Peres, H.E.M.; Sun Kou, R.; Gonçalves, J.M. Sensors based on Ag-loaded hematite (α-Fe_2O_3) nanoparticles for methyl mercaptan detection at room temperature. *Anal. Chem. Res.* **2017**, *12*, 74–81. [CrossRef]
42. Zhou, B.; Jiang, Y.; Guo, Q.; Das, A.; Sobrido, A.B.J.; Hing, K.A.; Zayats, A.V.; Krause, S. Photoelectrochemical Detection of Calcium Ions Based on Hematite Nanorod Sensors. *ACS Appl. Nano Mater.* **2022**, *5*, 17087–17094. [CrossRef] [PubMed]
43. Lee, S.; Jang, H.W. α-Fe2O3 nanostructure-based gas sensors. *J. Sens. Sci. Technol.* **2021**, *30*, 210–217. [CrossRef]
44. Tulliani, J.-M.; Baroni, C.; Zavattaro, L.; Grignani, C. Strontium-Doped Hematite as a Possible Humidity Sensing Material for Soil Water Content Determination. *Sensors* **2013**, *13*, 12070–12092. [CrossRef] [PubMed]
45. Hjiri, M. Highly sensitive NO_2 gas sensor based on hematite nanoparticles synthesized by sol–gel technique. *J. Mater. Sci. Mater. Electron.* **2020**, *31*, 5025–5031. [CrossRef]
46. Shen, S.; Lindley, S.A.; Chen, X.; Zhang, J.Z. Hematite heterostructures for photoelectrochemical water splitting: Rational materials design and charge carrier dynamics. *Energy Environ. Sci.* **2016**, *9*, 2744–2775. [CrossRef]
47. Park, J.; Kang, J.; Chaule, S.; Jang, J.-H. Recent progress and perspectives on heteroatom doping of hematite photoanodes for photoelectrochemical water splitting. *J. Mater. Chem. A* **2023**, *11*, 24551–24565. [CrossRef]
48. Zhu, L.; Li, Z.; Cheng, Y.; Zhang, X.; Du, H.; Zhu, C.; Jiang, D.; Yuan, Y. A highly efficient hematite photoelectrochemical fuel cell for solar-driven hydrogen production. *Int. J. Hydrog. Energy* **2023**, *48*, 32699–32707. [CrossRef]
49. Klahr, B.; Gimenez, S.; Fabregat-Santiago, F.; Bisquert, J.; Hamann, T.W. Electrochemical and photoelectrochemical investigation of water oxidation with hematite electrodes. *Energy Environ. Sci.* **2012**, *5*, 7626–7636. [CrossRef]

50. Wang, D.; Chang, G.; Zhang, Y.; Chao, J.; Yang, J.; Su, S.; Wang, L.; Fan, C.; Wang, L. Hierarchical three-dimensional branched hematite nanorod arrays with enhanced mid-visible light absorption for high-efficiency photoelectrochemical water splitting. *Nanoscale* **2016**, *8*, 12697–12701. [CrossRef]
51. Zhang, G.; Zhang, X.; Huang, H.; Wang, J.; Li, Q.; Chen, L.Q.; Wang, Q. Toward Wearable Cooling Devices: Highly Flexible Electrocaloric Ba0.67Sr0.33TiO3 Nanowire Arrays. *Adv. Mater.* **2016**, *28*, 4811–4816. [CrossRef] [PubMed]
52. Steier, L.; Luo, J.; Schreier, M.; Mayer, M.T.; Sajavaara, T.; Grätzel, M. Low-Temperature Atomic Layer Deposition of Crystalline and Photoactive Ultrathin Hematite Films for Solar Water Splitting. *ACS Nano* **2015**, *9*, 11775–11783. [CrossRef]
53. Dobbelaere, T.; Mattelaer, F.; Roy, A.K.; Vereecken, P.; Detavernier, C. Plasma-enhanced atomic layer deposition of titanium phosphate as an electrode for lithium-ion batteries. *J. Mater. Chem. A* **2017**, *5*, 330–338. [CrossRef]
54. Henderick, L.; Blomme, R.; Minjauw, M.; Keukelier, J.; Meersschaut, J.; Dendooven, J.; Vereecken, P.; Detavernier, C. Plasma-enhanced atomic layer deposition of nickel and cobalt phosphate for lithium ion batteries. *Dalton Trans.* **2022**, *51*, 2059–2067. [CrossRef]
55. Wu, W.-Q.; Lei, B.-X.; Rao, H.-S.; Xu, Y.-F.; Wang, Y.-F.; Su, C.-Y.; Kuang, D.-B. Hydrothermal Fabrication of Hierarchically Anatase TiO_2 Nanowire arrays on FTO Glass for Dye-sensitized Solar Cells. *Sci. Rep.* **2013**, *3*, 1352. [CrossRef] [PubMed]
56. Harris-Lee, T.R.; Zhang, Y.; Bowen, C.R.; Fletcher, P.J.; Zhao, Y.; Guo, Z.; Innocent, J.W.F.; Johnson, S.A.L.; Marken, F. Photo-Chlorine Production with Hydrothermally Grown and Vacuum-Annealed Nanocrystalline Rutile. *Electrocatalysis* **2021**, *12*, 65–77. [CrossRef]
57. Harris-Lee, T.R.; Johnson, S.A.L.; Wang, L.; Fletcher, P.J.; Zhang, J.; Bentley, C.; Bowen, C.R.; Marken, F. TiO_2 nanocrystal rods on titanium microwires: Growth, vacuum annealing, and photoelectrochemical oxygen evolution. *New J. Chem.* **2022**, *46*, 8385–8392. [CrossRef]
58. Lianos, P. Review of recent trends in photoelectrocatalytic conversion of solar energy to electricity and hydrogen. *Appl. Catal. B Environ.* **2017**, *210*, 235–254. [CrossRef]
59. Liu, J.; Yang, S.; Wu, W.; Tian, Q.; Cui, S.; Dai, Z.; Ren, F.; Xiao, X.; Jiang, C. 3D Flowerlike α-Fe_2O_3@TiO_2 Core–Shell Nanostructures: General Synthesis and Enhanced Photocatalytic Performance. *ACS Sustain. Chem. Eng.* **2015**, *3*, 2975–2984. [CrossRef]
60. Peng, L.; Xie, T.; Lu, Y.; Fan, H.; Wang, D. Synthesis, photoelectric properties and photocatalytic activity of the Fe_2O_3/TiO_2 heterogeneous photocatalysts. *Phys. Chem. Chem. Phys.* **2010**, *12*, 8033–8041. [CrossRef] [PubMed]

crystals

MDPI

Article

Deposition and Structural Characterization of Mg-Zn Co-Doped GaN Films by Radio-Frequency Magnetron Sputtering in a N_2-Ar_2 Environment

Erick Gastellóu [1,2,*], Rafael García [2], Ana M. Herrera [2], Antonio Ramos [2], Godofredo García [3], Gustavo A. Hirata [4], José A. Luna [3], Jorge A. Rodríguez [1], Mario Robles [1], Yani D. Ramírez [5] and Iván E. García [3]

1 División de Sistemas Automotrices, Universidad Tecnológica de Puebla (UTP), Antiguo Camino a la Resurrección 1002-A, Zona Industrial, Puebla 72300, Puebla, Mexico; jorge.rodriguez@utpuebla.edu.mx (J.A.R.); mario.robles@utpuebla.edu.mx (M.R.)
2 Departamento de Investigación en Física, Universidad de Sonora (UNISON), Rosales y Colosio, C. De la Sabiduría, Centro, Hermosillo 83000, Sonora, Mexico; rafael.gutierrez@unison.mx (R.G.); ana.herrera@unison.mx (A.M.H.); antonio.ramos@unison.mx (A.R.)
3 Centro de Investigacion en Dispositivos Semiconductores, Benemérita Universidad Autónoma de Puebla (BUAP), 14 Sur y Av. San Claudio, Puebla 72570, Puebla, Mexico; godofredo.garcia@correo.buap.mx (G.G.); jose.luna@correo.buap.mx (J.A.L.); ivan.garciabal@alumno.buap.mx (I.E.G.)
4 Centro de Nanociencias y Nanotecnología, Universidad Nacional Autónoma de Mexico (UNAM), Carr. Tijuana-Ensenada km 107, C.I.C.E.S.E., Ensenada 22860, Baja California, Mexico; hirata@ens.cnyn.unam.mx
5 Departamento de Investigación y Desarrollo, Universidad Tecnológica de Puebla (UTP), Antiguo Camino a La Resurrección 1002-A, Zona Industrial, Puebla 72300, Puebla, Mexico; yani.ramirez@utpuebla.edu.mx
* Correspondence: erick_gastellou@utpuebla.edu.mx; Tel.: +52-222-469-9594

Abstract: Mg-Zn co-doped GaN films were deposited by radio-frequency magnetron sputtering in an N_2-Ar_2 environment at room temperature, using a target prepared with Mg-Zn co-doped GaN powders. X-ray diffraction patterns showed broad peaks with an average crystal size of 13.65 nm and lattice constants for a hexagonal structure of a = 3.1 Å and c = 5.1 Å. Scanning electron microscopy micrographs and atomic force microscopy images demonstrated homogeneity in the deposition of the films and good surface morphology with a mean roughness of 1.1 nm. Energy-dispersive spectroscopy and X-ray photoelectron spectroscopy characterizations showed the presence of gallium and nitrogen as elemental contributions as well as of zinc and magnesium as co-doping elements. Profilometry showed a value of 260.2 nm in thickness in the Mg-Zn co-doped GaN films. Finally, photoluminescence demonstrated fundamental energy emission located at 2.8 eV (430.5 nm), which might be related to the incorporation of magnesium and zinc atoms.

Keywords: co-doped; GaN; film; radio-frequency magnetron sputtering

Citation: Gastellóu, E.; García, R.; Herrera, A.M.; Ramos, A.; García, G.; Hirata, G.A.; Luna, J.A.; Rodríguez, J.A.; Robles, M.; Ramírez, Y.D.; et al. Deposition and Structural Characterization of Mg-Zn Co-Doped GaN Films by Radio-Frequency Magnetron Sputtering in a N_2-Ar_2 Environment. *Crystals* **2024**, *14*, 618. https://doi.org/10.3390/cryst14070618

Academic Editors: Dah-Shyang Tsai and Evgeniy N. Mokhov

Received: 31 May 2024
Revised: 21 June 2024
Accepted: 2 July 2024
Published: 4 July 2024

1. Introduction

Nowadays, the global shortage of semiconductors has impacted different industries, such as electrotechnology, automotive, and biomedicine [1]. Currently, one of the main semiconductors is GaN, which belongs to the III-Nitride family. This semiconductor is very widely used due to its wide band gap of 3.4 eV for a wurtzite structure or 3.2 eV for a zincblende crystalline structure [2,3]. GaN has applications in solar cells, LED screens, LED technology, high-electron-mobility transistors (HEMTs), microwave devices, photocatalysis, and laser diodes [2–7].The doping and co-doping of GaN have attracted the interest of researchers due to the fact that these techniques can vary in their structural, optical, and electrical properties. Some works have investigated the process of obtaining GaN co-doped with different doping elements.

Liu et al. presented the obtaining of Si-Ti co-doped GaN films via sputtering with a zinc oxide (ZnO) buffer layer on amorphous glass substrates, where the n-type films had a resistivity of 2.6×10^{-1} Ω-cm [8]. Sun et al. fabricated Sm-Eu co-doped GaN films using co-implantation of ions into a c-plane, and after conducting an annealing process, they studied the structural, morphological, and magnetic characteristics of the films [9]. Jeong et al. grew Mg-Mn co-doped films with low Mg and Mn. concentrations using plasma-enhanced molecular beam epitaxy (PEMBE), and the samples showed n-type conductivity and ferromagnetism at room temperature [10]. Kim et al. grew Mg-Si co-doped GaN films using metalorganic chemical vapor deposition (MOCVD), and high p-type conductivity was obtained, besides competitive adsorption between Mg and Si during the growth [11]. In another work, Kim et al. showed the doping characteristics of Mg-Zn co-doped GaN films grown using metalorganic chemical vapor deposition (MOCVD), where a p-type conductivity with a hole concentration of 8.5×10^{17} cm^{-3} was obtained. It is important to mention that in this last work, only electric characteristics were studied [12]. Naito et al. presented the epitaxial growth of In-Mg co-doped GaN with a hole concentration of 6.2×10^{18} cm^{-3} without structural degradation via pulsed sputtering deposition (PSD) [13].

In a previous work, our research team synthesized Mg-Zn co-doped GaN powders via nitridation of the Ga-Mg-Zn metallic liquid solution at 1000 °C for two hours. In that research study, the metallic liquid solution was homogenized above the melting temperatures of the Mg and Zn to supersaturate the liquid metal solution, and a nitridation process was subsequently performed [14]. In another previous work, Mg-doped GaN powders and Zn-doped GaN powders were used as raw materials to prepare targets in the laboratory. Afterwards, a tableting process was used to prepare the targets, which underwent a sintering process. Once the targets were obtained, they were used for growth via radio-frequency magnetron sputtering of Mg-doped GaN films and Zn-doped GaN films in a N_2 environment. The films were characterized by different structural, optical, and electrical techniques [15].

The aim of this work is to obtain and conduct a structural analysis of Mg-Zn co-doped GaN films deposited by radio-frequency magnetron sputtering in an N_2-Ar_2 environment using targets prepared with Mg-Zn co-doped GaN powders. In this study, the growth environment in N_2-Ar_2 plasma is expected to improve the crystalline quality of the Mg-Zn co-doped GaN films, and these Mg-Zn co-doped GaN films could have application in SARS-CoV-2 biosensors due to the fact that GaN is biocompatible and non-toxic with the functionalization of peptides [16]. The structural properties of the Mg-Zn co-doped GaN films were characterized via X-ray diffraction patterns (XRD); besides this, their surface morphology and topography were measured by scanning electron microscopy (SEM) and atomic force microscopy (AFM). On the other hand, the elemental contributions were characterized using energy-dispersive spectroscopy (EDS) and X-ray photoelectron spectroscopy (XPS). To measure the thickness of the films, profilometry was used, and to measure the energy emission of the Mg-Zn co-doped GaN films, the photoluminescence (PL) technique was used.

2. Experimental

2.1. Preparation of the Raw Material to Prepare the Mg-Zn Co-Doped GaN Target

To obtain the Mg-Zn co-doped GaN powders, or raw material to prepare the Mg-Zn co-doped GaN target, the following method was used: A simple process was carried out at atmospheric pressure (1010 hPa). Ultra-high-purity metallic gallium from Sigma-Aldrich (St. Louis, MI, USA) (1.3 g), ultra-high-purity metallic zinc (7.8 mg), and ultra-high-purity metallic magnesium from Sigma-Aldrich (5.2 mg) were used as reagents, and ultra-high-purity ammonia was used as a source of nitrogen atoms. The reagents were placed in a high-alumina boat, which was placed on a hot plate at 200 °C for manual stirring for two hours to mix the liquid materials. Later, the boat along with the reagents were placed inside a CVD furnace (chemical vapor deposition), which was purged three times with a flow of ultra-high-purity nitrogen at 50 sccm to eliminate oxygen and other residual contaminants After, nitrogen flowed at 100 sccm as a carrier gas, and the system was programmed at

440 °C for 40 min in a nitrogen environment to supersaturate the zinc, 20 °C above its melting temperature. Immediately after, the CVD system was reprogrammed, now at 670 °C for 40 min in a nitrogen flow at 100 sccm; this is also 20 °C above the melting temperature of magnesium. With this process the supersaturation of the Ga-Mg-Zn liquid metal solution was achieved. Afterward, the temperature was raised to 900 °C for 40 min, where the nitrogen flow was closed and a flow of ammonia (NH_3) was opened at 100 sccm to prepare the nitridation process, during which the temperature was raised to 1000 °C for two hours in a NH_3 environment. Once this process was completed, the temperature was lowered to 400 °C, the ammonia flow was closed, and the nitrogen flow was opened again at 100 sccm until until it reached room temperature. At the end of this process, the boat with the raw material was extracted from the CVD system and was prepared for grinding [14].

2.2. Preparation of the Mg-Zn Co-Doped GaN Target for Film Deposition

Once the Mg-Zn co-doped GaN powders were synthesized, the tableting process began to obtain the targets for the deposition of the films via radio-frequency magnetron sputtering. The general procedure for the preparation of the targets is described as follows: A very fine grinding of the Mg-Zn co-doped GaN powders was conducted using an agate mortar to homogenize the size of the powder before compaction. After the finer grinding was carried out, the Mg-Zn co-doped GaN powders were lubricated using 0.5 mL of ultra-high-purity ethanol. The target mold was then placed in a 25-ton Blackhawk SP25B press. The Mg-Zn co-doped GaN powders were placed inside the target mold and pressed at approximately 10 tons per cm^2. Once the powders were compacted, the Mg-Zn co-doped GaN target was extracted; then, the Mg-Zn co-doped GaN target was individually sintered at 900 °C for one hour inside a CVD furnace in a flow of NH_3 at 100 sccm. Afterwards, the CVD furnace was programmed to decrease its temperature to 400 °C; at this temperature, the NH_3 flow was closed, and a nitrogen flow was opened at 100 sccm. Then, the temperature continued to decrease until reaching room temperature in a nitrogen flow at 100 sccm. When room temperature was reached, the CVD system was purged, and the vacuum was subsequently broken. After the target was extracted and its hardness tested, the procedure was carried out again, from grinding, tableting, and sintering. If the target had the necessary hardness for deposition by sputtering, the process of obtaining the Mg-Zn co-doped GaN target was complete [15].

2.3. Deposition of the Mg-Zn Co-Doped GaN Films in a N_2-Ar_2 Environment

The Mg-Zn co-doped GaN film was deposited by radio-frequency magnetron sputtering in a N_2-Ar_2 environment on silicon substrates (100) at room temperature (atmosphere) using targets prepared with Mg-Zn GaN powders, which were synthesized as mentioned in our previous work (Gastellóu et al.) [14]. Therefore, using the process in reference [14], 11.1652 g of Mg-Zn GaN powders was synthesized, with a percentage of 0.4% of magnesium and 0.6% of zinc. Once synthesized, the Mg-Zn GaN powders were prepared for the Mg-Zn co-doped GaN targets following the process presented in another of our previous works (Gastellou et al.) [15]. It is important to mention that the target was 50.8 mm in diameter, while the thickness was 5 mm. Once the Mg-Zn co-doped GaN target was obtained, the silicon (100) substrates were cleaned with a conventional process of solvents and solutions to remove organic residues, and after that they were placed inside a beak with methanol to prepare their introduction into the sputtering chamber. The Mg-Zn co-doped GaN film was deposited using an Intercovamex Sputtering System V1 with the following conditions: a separation distance of 40 mm was added between the substrate and the target (modification made in the system for this work); the chamber vacuum attained a pressure of 2×10^{-6} Torr before the film growth. N_2-Ar_2 (50–50%) flows were used during the sputtering process; besides an RF power of 60 W, a gas pressure of 15×10^{-3} Torr was kept during the film deposition, which generated a violet plasma. The growth time was 3 h. To deposit the Mg-Zn co-doped GaN films, the target was placed in the sputtering magnetron, which has a smooth surface to correctly couple the target and allow for its cooling with the equipment's system; subsequently, the target was fixed to the magnetron using the "cap". Afterwards, the substrate holder was placed 4 cm from the surface of the sputtering magnetron, the chamber was closed, and the sputtering system was turned on. It is important to mention that during the deposition of the Mg-Zn co-doped GaN films, the internal chamber sensor indicated a maximum temperature of 41 °C; this is due to the uniform cooling system of the sputtering system itself (surrounding the chamber). Once the deposition was finished, the cooling of the target and the substrate holder was carried out in conjunction with the shutdown of the sputtering system and the cooling system, which took approximately one hour, before opening the chamber to be able to extract the substrate holder with the deposited films and the target. Figure 1 shows the violet plasma produced by the mixture of nitrogen and argon generated during the deposition of the Mg-Zn co-doped GaN films.

Figure 1. Violet plasma is generated during the deposition of the Mg-Zn co-doped GaN films.

2.4. Characterizations

Mg-Zn co-doped GaN films deposited via radio-frequency magnetron sputtering in a N_2-Ar_2 environment were characterized by X-ray diffraction patterns (XRD) using Bruker AXS D8 Discover equipment (Bruker, Karlsruhe, Germany) with a wavelength (Cu Kα) of 1.5406 Å, in a range from 20 to 60 degrees, and grazing incidence X-ray diffraction (GIXRD). Profilometry to measure the thickness of the films was conducted using a Dektak 150 Surface Profiler (Veeco, Tucson, AZ, USA). Surface morphology and elemental contributions (SEM-EDS) were measured using JEOL JSM-7800F Schottky Field Emission equipment (JEOL, Pleasanton, CA, USA). The analysis of the surface topography of the Mg-Zn co-doped GaN films was performed using a Veeco Nanoscope IIIa Atomic Force Microscope (AFM) (Veeco, Tucson, AZ, USA). X-ray photoelectron spectroscopy (XPS) measurements were carried out using Escalab 250Xi Brochure equipment (Thermo Scientific, East Grinstead, UK) with an energy range from 0 to 12 KeV. Finally, the photoluminescence spectrum (PL) was realized at room temperature with a fluorescence spectrophotometer Hitachi F-7000 FL (Hitachi, Tokyo, Japan) with an excitation wavelength of 243 nm and a 310 nm filter with a 150 W xenon lamp.

3. Results and Discussion

3.1. Structural Analysis

Mg-Zn co-doped GaN films were deposited on silicon substrates via radio-frequency magnetron sputtering in a N_2-Ar_2 environment. Figure 2 shows the X-ray diffraction patterns of Mg-Zn co-doped GaN films (Figure 2a), which were indexed in the ICDD PDF card No. 00-050-0792 (Figure 2b). In Figure 2a, the *a* peak belongs in the (100) plane, the *b* peak is located in the (002) plane, the *c* peak belongs in the (101) plane, the *d* peak is located in the (102) plane, and the *e* peak belongs in the (110) plane. The lattice constants were calculated using the ICDD PDF card No. 00-050-0792 for the hexagonal structure, with a = 3.1 Å and c = 5.1 Å for the space group P63mc and with a c/a ratio of 1.6. The X-ray diffraction patterns in Figure 2a show an FWHM of 0.54° for the *a* peak, with a crystal size of 15.9 nm and an interplanar spacing of 2.7 Å. The FWHM measure for the *b* peak had a value of 0.9°, with a crystal size of 9.0 nm and an interplanar spacing of 2.6 Å. The *c* peak had an FWHM of 0.67°, with a crystal size of 12.9 nm and an interplanar spacing of 2.4 Å, while the *d* peak had an FWHM of 0.71°, with a crystal size of 12.7 nm and an interplanar spacing of 1.9 Å. Finally, the *e* peak had an FWHM of 0.53°, with a crystal size of 17.6 nm and an interplanar spacing of 1.6 Å. It is important to mention that the crystallite size of each peak was calculated using the Scherrer equation, shown below:

$$D = \frac{K\lambda}{\beta \cos\theta} \quad (1)$$

where D is the average crystallite size (nm), K is the Scherrer constant (0.94), λ is the X-ray wavelength (Cu Kα = 1.5 Å), β is the line broadening at FWHM in radians, and θ is Bragg's angle in degrees, half of 2θ. Furthermore, using the ICCD PDF-4+2022 software and the Debye–Scherrer equation, the average crystal size was calculated, and a value of 13.6 nm was found, which could indicate that the widening of the peaks in Figure 2a might be associated with the presence of nanocrystallites, as demonstrated by Garcia et al. [17]. The more defined XRD peaks shown in this work would be related to the environment of nitrogen with argon (50–50%) for the Mg-Zn co-doped GaN films, which contrasts with the XRD peaks in a nitrogen environment (100%) presented in our previous work [16].

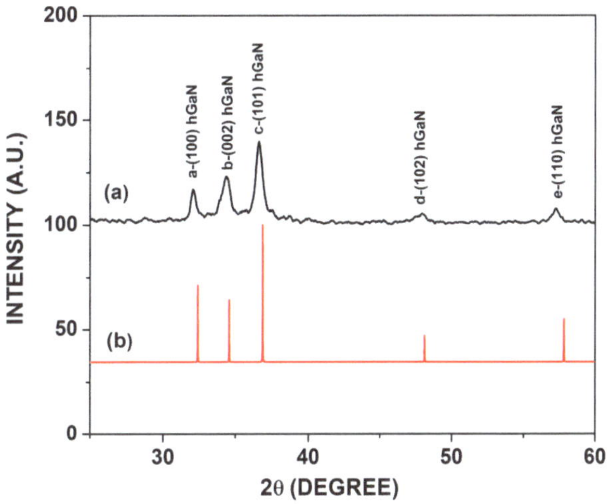

Figure 2. X-ray diffraction patterns: (**a**) Mg-Zn co-doped GaN films deposited by radio-frequency magnetron sputtering in N_2-Ar_2 environment; (**b**) ICDD PDF card No. 00-050-0792.

3.2. Electron Microscopy

Figure 3 shows the SEM micrographs for the Mg-Zn co-doped GaN films. Figure 3a demonstrates a surface morphology where the deposition via radio-frequency magnetron sputtering generated good cohesion between the silicon substrate and the film with a magnification of X5000. Furthermore, a good uniform morphology can be observed in Figure 3a. Figure 3b also shows good homogeneity in the deposition of the Mg-Zn co-doped GaN films with a magnification of X50,000; however, little agglomerates could be observed on the surface, which might be related to structural defects such as oxygen interstitial.

This oxygen interstitial could be formed during the hysteresis effect due to growth via radio-frequency magnetron sputtering [16].

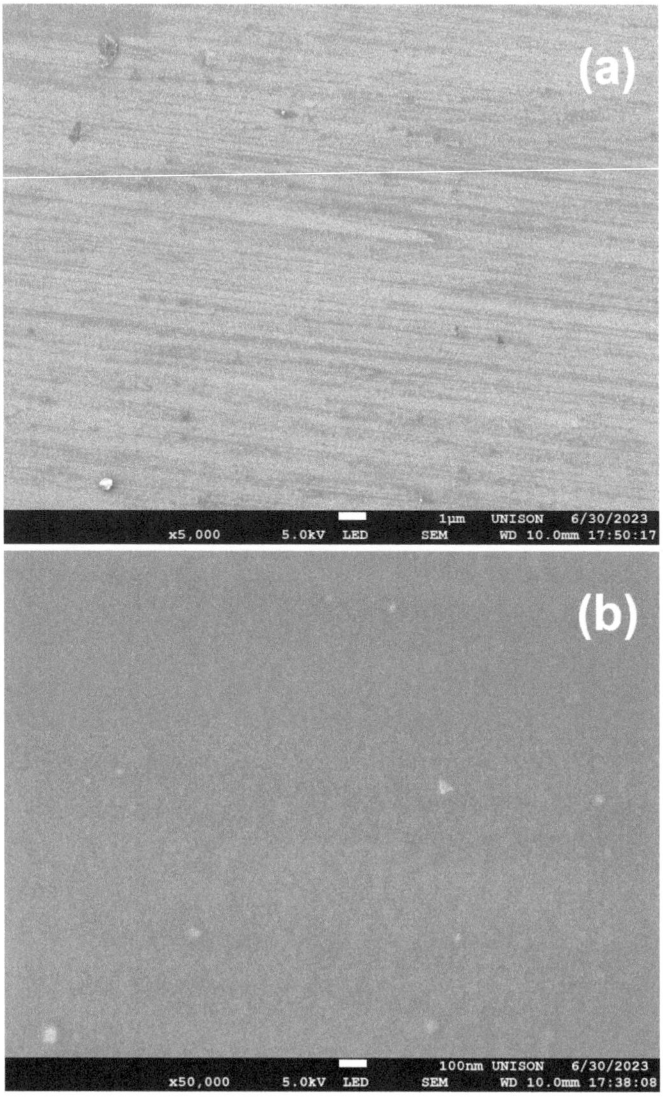

Figure 3. SEM micrographs of Mg-Zn co-doped GaN films with magnifications of (**a**) X5000, and (**b**) X50,000.

Figure 4 shows the EDS spectrum of the elemental analysis of the Mg-Zn co-doped GaN films, where elemental contributions of gallium were found at Lα-1.098 eV and Kα-9.2 eV. In addition, the elemental contribution of nitrogen was found at Kα-0.3 eV, and the elemental contribution of magnesium was found at Kα-1.2 eV. Finally, the elemental contribution of zinc was found at Lα-1.0 eV. The atomic percentages in the EDS analysis were gallium (49.66%), nitrogen (31.81%), zinc (0.60%), magnesium (0.32%), oxygen (1.87%), and carbon (15.73%). Furthermore, located at 8 eV, the presence of a small copper signal belonging to the sample holder of the EDS system is shown. To find the thickness of the Mg-Zn co-doped GaN films deposited by radio-frequency magnetron sputtering in a

N_2-Ar_2 environment, which had a value of 260.2 nm, the technique of profilometry was used. The EDS analysis of the Mg-Zn co-doped GaN powders with which the target for the deposition of the films was prepared was carried out in ref. [14].

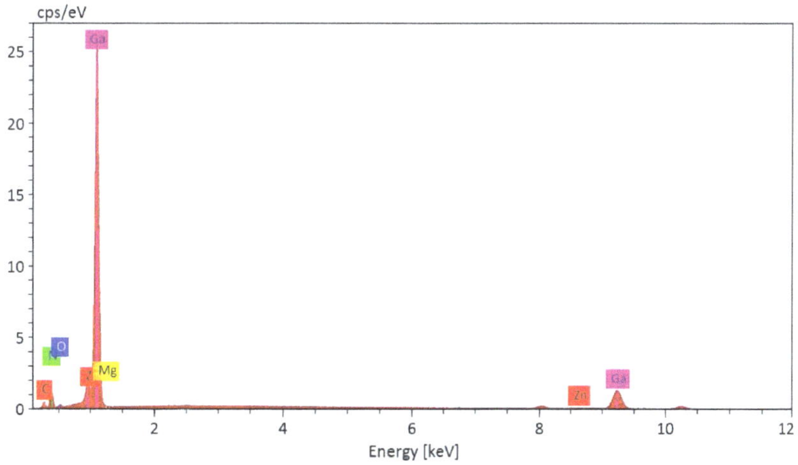

Figure 4. EDS analysis of Mg-Zn co-doped GaN films deposited by radio-frequency magnetron sputtering in N_2-Ar_2 environment.

Figure 5 shows the analysis of surface topography using the technique of atomic force microscopy (AFM). Figure 5 illustrates 2D (Figure 5a) and 3D (Figure 5b) topographical AFM images of the Mg-Zn co-doped GaN films. The AFM images were measured over an area of 5 × 5 μm^2. In Figure 5, using the software Gwyddion 2.63, a mean roughness of 1.1 nm was calculated, while the root mean square roughness was 1.3 nm and the maximum peak height was 5.0 nm. These AFM images demonstrate good surface morphology, besides good homogeneity, which agrees with the SEM micrographs of the deposition of the Mg-Zn co-doped GaN films [9,18,19].

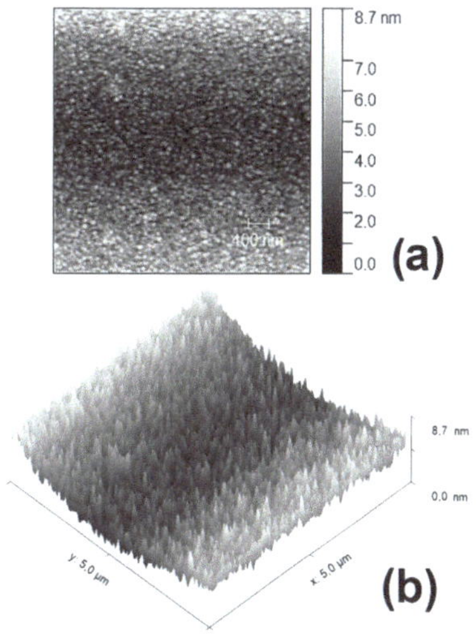

Figure 5. (**a**) Two-dimensional and (**b**) three-dimensional AFM images of Mg-Zn co-doped GaN films deposited by radio-frequency magnetron sputtering in N_2-Ar_2 environment.

3.3. X-ray Photoelectron Spectroscopy

Figure 6 shows the XPS spectra of the Mg-Zn co-doped GaN films deposited by radio-frequency magnetron sputtering in a N_2-Ar_2 environment. Figure 6a presents the peaks for high energies of Ga $2P_{1/2}$ and Ga $2P_{3/2}$, with values of 1146.3 eV and 1119.7 eV, respectively. Figure 6b presents the N 1s peak with an energy value of 399.2 eV. Figure 6c presents the Zn $2P_{3/2}$ peak with an energy value of 1019.4 eV, while Figure 6d shows the Mg $2P_{3/2}$ peak with a binding energy of 49.7 eV. These results agree with those found in the EDS analysis, due to the presence of the elemental contributions of gallium, nitrogen, magnesium, and zinc.

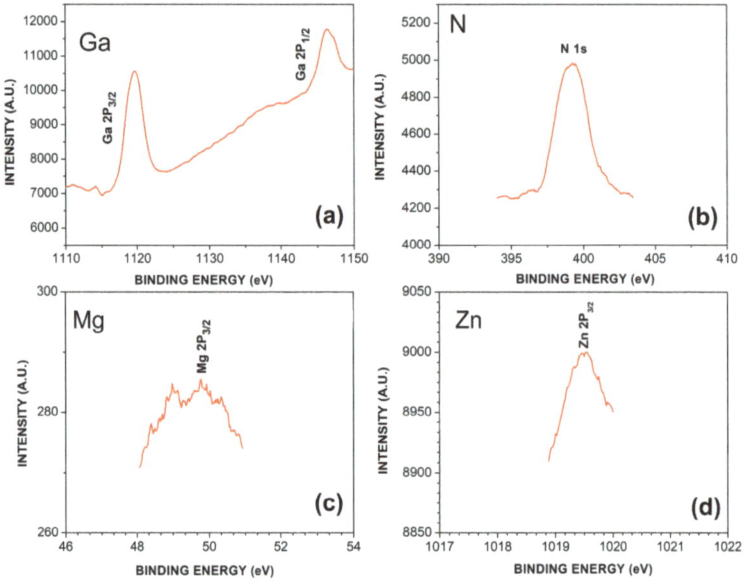

Figure 6. XPS spectra for Mg-Zn co-doped GaN films deposited by radio-frequency magnetron sputtering in N_2-Ar_2 environment: (**a**) Ga $2P_{3/2}$ and Ga $2P_{1/2}$, (**b**) N 1s, (**c**) Mg $2P_{3/2}$, (**d**) Zn $2P_{3/2}$.

3.4. Photoluminescence

Figure 7 shows the photoluminescence spectrum for the Mg-Zn co-doped GaN films deposited by radio-frequency magnetron sputtering in a N_2-Ar_2 environment. Figure 7 demonstrates four energy emissions. The **a (Mg-Zn co-doped GaN)** energy emission corresponds to 2.8 eV (430.5 nm) and could be related to the incorporation of magnesium and zinc atoms. The blue luminescence (BL) for the incorporation of zinc atoms into GaN has a value of 2.8 eV, while the blue luminescence (BL) for the incorporation of magnesium atoms into GaN has a value from 2.7 to 3.0 eV [20,21]. Therefore, this value of energy emission agrees with the incorporation of zinc and magnesium into GaN for obtaining the co-doping, besides agreeing with the EDS and XPS analyses. The **b (GaN:C)** energy emission is located at 2.2 eV (546.2 nm) and could be related to non-intentional impurities of carbon in yellow luminescence (YL). These carbon impurities might be due to a little elemental contribution that appears in the EDS analysis, besides impurities belonging to the Mg-Zn co-doped GaN powders with which the target was made, which were transferred to the film in the deposition by sputtering. The **c (GaN:C)** energy emission located at 2.1 eV (582.1 nm) is also related to non-intentional impurities of carbon, while the **d (GaN)** energy emission is located at 2.0 eV (620 nm) and belongs to the red luminescence (RL) of GaN.

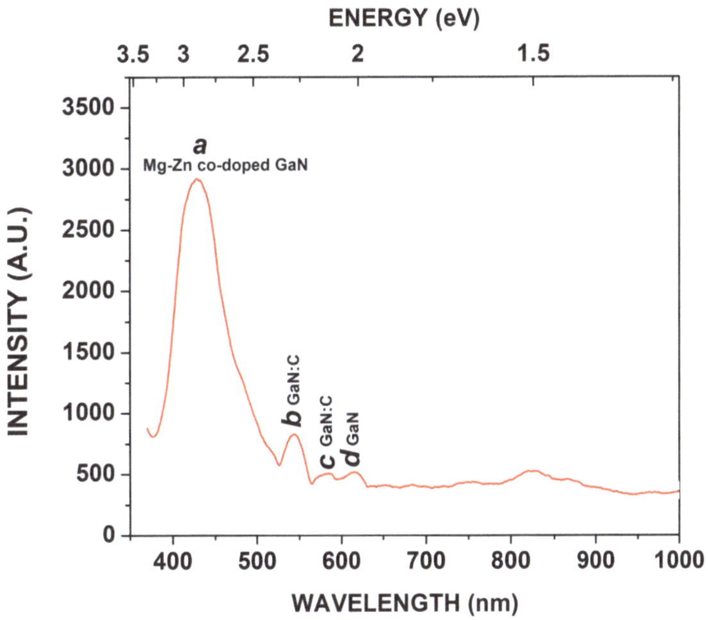

Figure 7. Photoluminescence spectrum of Mg-Zn co-doped GaN films deposited by radio-frequency magnetron sputtering in N_2-Ar_2 environment.

4. Conclusions

Mg-Zn co-doped GaN films were deposited by radio-frequency magnetron sputtering in a N_2-Ar_2 environment at room temperature using a target prepared with Mg-Zn co-doped GaN powders. The X-ray diffraction patterns (XRD) showed a higher intensity peak with an FWHM of 0.67°, a crystal size of 12.9 nm, an interplanar spacing of 2.4 Å, and an average crystal size of 13.6 nm. SEM micrographs showed good homogeneity in the deposition, which agrees with the AFM images with a mean roughness of 1.1 nm. The profilometry had a value of 260.2 nm for the thickness of the Mg-Zn co-doped GaN films. The EDS spectrum and the XPS analysis demonstrated the presence of gallium and nitrogen as elemental contributions, besides the co-doping with zinc and magnesium. Finally, the PL spectrum showed that the fundamental energy emission corresponds to 2.8 eV (430.5 nm) and could be related to the incorporation of magnesium and zinc atoms in the blue luminescence (BL). In addition, PL analysis also demonstrated carbon impurities belonging to the Mg-Zn co-doped GaN powders with which the target was made and that were transferred to the film in the deposition by sputtering, which agrees with the EDS analysis.

Author Contributions: Conceptualization, E.G.; Methodology, Formal Analysis, Investigation, Writing—Original Draft, Resources, R.G.; Formal Analysis, Investigation, Writing—Original Draft, A.M.H.; Resources, A.R.; Conceptualization, G.G.; Resources, G.A.H.; Resources, J.A.L.; Resources, J.A.R.; Resources, M.R.; Validation, Y.D.R.; Resources, I.E.G. All authors have read and agreed to the published version of the manuscript.

Funding: This research received no external funding.

Data Availability Statement: Data are contained within the article.

Acknowledgments: The authors gratefully acknowledge the technical support of CNyN-UNAM, DF-UNISON, and IFUAP-BUAP. This research was partially supported by CONAHCyT (Mexico), DIFUS-UNISON, and UTPuebla.

Conflicts of Interest: The authors declare no conflicts of interest.

References

1. Wassen, M.; Adel, E.; Laoucine, K. The Global Semiconductor Chip Shortage: Causes, Implications, and Potential Remedies. *IFAC-PapersOnLine* **2022**, *55*, 476–483. [CrossRef]
2. Kwon, W.; Kawasaki, S.; Watanabe, H.; Tanaka, A.; Honda, Y.; Ikeda, H.; Iso, K.; Amano, H. Reverse Leakage Mechanism of Dislocation-Free GaN Vertical p-n Diodes. *IEEE Electron Device Lett.* **2023**, *44*, 1172–1175. [CrossRef]
3. Fanciulli, M.; Lei, T.; Moustakas, T.D. Conduction-electron spin resonance in zinc-blende GaN thin films. *Phys. Rev. B* **1993**, *20*, 15-144–15-147. [CrossRef] [PubMed]
4. Sami, B.; Burak, T.; Cagla, O.-A.; Necmi, B.; Kemal, O.A. Electronic and optical device applications of hollow cathode plasma assisted atomic layer deposition based GaN thin films. *J. Vac. Sci. Technol. A Vac. Surf. Film.* **2015**, *33*, 01A143. [CrossRef]
5. Lee, D.S.; Lee, J.H.; Lee, Y.H.; Lee, D.D. GaN thin films as gas sensors. *Sens. Actuators B* **2003**, *89*, 305–310. [CrossRef]
6. Huang, S.; Zhang, Y.; Leung, B.; Yuan, G.; Wang, G.; Jiang, H.; Fan, Y.; Sun, Q.; Wang, J.; Xu, K.; et al. Mechanical Properties of Nanoporous GaN and Its Application for Separation and Transfer of GaN Thin Films. *ACS Appl. Mater. Interfaces* **2013**, *5*, 11074–11079. [CrossRef] [PubMed]
7. Jung, H.S.; Hong, Y.J.; Li, Y.; Cho, J.; Kim, Y.J.; Yi, G.C. Photocatalysis Using GaN Nanowires. *ACS Nano* **2008**, *2*, 637–642. [CrossRef]
8. Liu, W.-S.; Chang, Y.-L.; Tan, C.-Y.; Tsai, C.-T.; Kuo, H.-C. Properties of N-Type GaN Thin Film with Si-Ti Codoping on a Glass Substrate. *Crystals* **2020**, *10*, 582. [CrossRef]
9. Sun, L.; Liu, C.; Li, J.; Wang, J.; Yan, F.; Zeng, Y.; Li, J. Structural and magnetic properties of GaN:Sm:Eu films fabricated by co-implantation method. *Mater. Lett.* **2010**, *64*, 1031–1033. [CrossRef]
10. Jeong, M.C.; Ham, M.H.; Myoung, J.M.; Noh, S.K. Room-temperature ferromagnetism of Mg and Mn co-doped GaN films grown by PEMBE. *Appl. Surf. Sci.* **2004**, *222*, 322–326. [CrossRef]
11. Kim, K.S.; Yang, G.M.; Lee, H.J. The study on the growth and properties of Mg doped and Mg–Si codoped p-type GaN. *Solid-State Electron.* **1999**, *43*, 1807–1812. [CrossRef]
12. Kim, K.S.; Han, M.S.; Yang, G.M.; Youn, C.J.; Lee, H.J.; Cho, H.K.; Lee, J. Codoping characteristics of Zn with Mg in GaN. *Appl. Phys. Lett.* **2000**, *77*, 1123–1125. [CrossRef]
13. Naito, A.; Ueno, K.; Kobayashi, A.; Fujioka, H. Hole Conduction Mechanism in In–Mg-Codoped GaN Prepared via Pulsed Sputtering Deposition. *Phys. Status Solidi A* **2024**, *221*, 2300806. [CrossRef]
14. Gastellóu, E.; García, R.; Herrera, A.M.; Ramos, A.; García, G.; Hirata, G.A.; Luna, J.A.; Carrillo, R.C.; Rodríguez, J.A.; Robles, M.; et al. Obtaining of Mg-Zn Co-Doped GaN Powders via Nitridation of the Ga-Mg-Zn Metallic Solution and Their Structural and Optical Properties. *Materials* **2023**, *16*, 3272. [CrossRef]
15. Gastellóu, E.; García, G.; Herrera, A.M.; Morales, C.; García, R.; Hirata, G.A.; Rosendo, E.; Luna, J.A.; Robles, M.; Rodríguez, J.A.; et al. Effects in the Optical and Structural Properties Caused by Mg or Zn Doping of GaN Films Grown via Radio-Frequency Magnetron Sputtering Using Laboratory-Prepared Targets. *Appl. Sci.* **2021**, *11*, 6990. [CrossRef]
16. Jewett, S.A.; Makowski, M.S.; Andrews, B.; Manfra, M.J.; Ivanisevic, A. Gallium nitride is biocompatible and non-toxic before and after functionalization with peptides. *Acta Biomater.* **2012**, *8*, 728–733. [CrossRef] [PubMed]
17. Garcia, R.; Hirata, G.A.; Thomas, A.C.; Ponce, F.A. Structure and luminescence of nanocrystalline gallium nitride synthesized by a novel polymer pyrolysis route. *Opt. Mater.* **2006**, *29*, 19–23. [CrossRef]
18. Fang, J.; Yang, W.; Zhang, X.; Tian, A.; Lu, S.; Liu, J.; Yang, H. The Effect of Periodic Duty Cyclings in Metal-Modulated Epitaxy on GaN:Mg Film. *Materials* **2023**, *16*, 1730. [CrossRef]
19. Kumar, A.; Berg, M.; Wang, Q.; Salter, M.; Ramvall, P. Growth of p-type GaN—The role of oxygen in activation of Mg-doping. *Power Electron. Devices Compon.* **2023**, *5*, 100036. [CrossRef]
20. Reshchikov, M.A.; Morkoç, H. Luminescence properties of defects in GaN. *J. Appl. Phys.* **2005**, *97*, 061301-1–061301-95. [CrossRef]
21. Furqan, C.M.; Ho Jacob, Y.L.; Kwok, H.S. GaN thin film: Growth and Characterizations by Magnetron Sputtering. *Surf. Interfaces* **2021**, *26*, 101364. [CrossRef]

crystals

MDPI

Article

High-Temperature $(Cu,C)Ba_2Ca_3Cu_4O_y$ Superconducting Films with Large Irreversible Fields Grown on $SrLaAlO_4$ Substrates by Pulsed Laser Deposition

Yugang Li, Zhiyong Liu *, Ping Zhu, Jinyu He and Chuanbing Cai *

Shanghai Key Laboratory of High Temperature Superconductors, Department of Physics, Shanghai University, Shanghai 200444, China; hejinyu@shu.edu.cn (J.H.)
* Correspondence: zyliu@shu.edu.cn (Z.L.); cbcai@shu.edu.cn (C.C.)

Abstract: $(Cu,C)Ba_2Ca_3Cu_4O_y$ is a nontoxic cuprate superconducting material with a superconducting transition temperature of about 116 K. Recently, it was found that bulk samples of this material synthesized under high pressure hold the highest irreversibility line among all the superconductors, which is very promising for its application in the liquid nitrogen temperature field. In this work, high-temperature $(Cu,C)Ba_2Ca_3Cu_4O_y$ superconducting films with large irreversible fields were prepared on $SrLaAlO_4(00l)$ substrates by pulsed laser deposition. The substrate temperature during deposition proved to be the most important parameter determining the morphology and critical temperature of the superconductors, with 680 °C considered to be the optimum temperature. X-ray diffraction (XRD) results showed that the $(Cu,C)Ba_2Ca_3Cu_4O_y$ films prepared under optimal conditions exhibited epitaxial growth with the *a*-axis perpendicular to the film surface and the *b*- and *c*-axes parallel to the substrate, with no evidence of any other orientation. In addition, resistivity measurements showed that the onset transition temperature (T_c^{onset}) was approximately 116 K, the zero-resistance critical temperature (T_{c0}) was around 53 K, and the irreversible field (H_{irr}) was about 9 T at 37 K for $(Cu,C)Ba_2Ca_3Cu_4O_y$ films under optimal temperature. This is the first example of the successful growth of superconducting $(Cu,C)Ba_2Ca_3Cu_4O_y$ films on $SrLaAlO_4(00l)$ substrates. This will facilitate high-performance applications of $(Cu,C)Ba_2Ca_3Cu_4O_y$ superconducting materials in the liquid nitrogen temperature field.

Keywords: $(Cu,C)Ba_2Ca_3Cu_4O_y$ thin film; $SrLaAlO_4$ substrate; pulsed laser deposition; substrate temperature

Citation: Li, Y.; Liu, Z.; Zhu, P.; He, J.; Cai, C. High-Temperature $(Cu,C)Ba_2Ca_3Cu_4O_y$ Superconducting Films with Large Irreversible Fields Grown on $SrLaAlO_4$ Substrates by Pulsed Laser Deposition. *Crystals* **2024**, *14*, 514. https://doi.org/10.3390/cryst14060514

Academic Editor: Serguei Petrovich Palto

Received: 9 May 2024
Revised: 21 May 2024
Accepted: 26 May 2024
Published: 28 May 2024

1. Introduction

Based on their superconducting transition temperature (T_c), superconductors can be divided into two categories: Low Temperature Superconductors (LTSs) and High Temperature Superconductors (HTSs). HTSs consist mainly of copper oxide superconductors, iron-based superconductors, and hydride superconductors. Copper oxide superconductors mainly include Tl-Ba-Ca-Cu-O, Hg-Ba-Ca-Cu-O, Bi-Sr-Ca-Cu-O, and RE-Ba-Cu-O [1–4]. The applications of Tl and Hg systems are limited by their toxic elements and the complexity of the manufacturing process. In contrast, the Bi and Y systems have produced three practical high-temperature superconducting material systems: Bi-2212, Bi-2223, and RE-123. The first generation of high-temperature superconducting tapes (1G-HTSs) is based on powder packaging and wire-drawing processes and consists of Bi-2212 and Bi-2223 wires. Bi-2223 has a T_c exceeding 100 K, but its very layered structure and huge anisotropy do not allow a high irreversibility field at the liquid nitrogen temperature [5,6]. In contrast, Bi-2212 has some limitations, such as low critical current density, high void ratio, poor mechanical properties, high production cost, etc., which limit its application range and development prospects. REBCO-coated conductors based on the development of film epitaxy and biaxial texture growth on flexible metal substrates are called second-generation high-temperature superconducting tapes (2G-HTSs) [7]. Their biaxial fabric structures eliminate weak-linking at grain boundaries and the defects such as dislocations and vacancies associated with the fabrication of superconducting layers are flux-pinning centres. As a result, the critical current density of REBCO-coated conductors in the liquid nitrogen temperature range is significantly higher than that of other materials, as studies have shown [8–10]. However, the T_c of REBCO superconductors is only about 90 K. in contrast, another cuprate superconductor without toxic elements, (Cu,C)$Ba_2Ca_3Cu_4O_y$ ((Cu,C)-1234), shows structural similarities to $HgBa_2Ca_3Cu_4O_y$ [11], with a T_c of about 116 K. At a field of zero, the critical current density J_c reaches 6×10^6 A cm^{-2} at 4.2 K and 6.5×10^5 A cm^{-2} at 77 K [12]. Among superconducting materials to date, polycrystalline (Cu,C)-1234 blocks exhibit the highest H_{irr} in the temperature range of liquid nitrogen [12], reaching 15 T at 85 K and 5 T at 98 K, respectively. These excellent properties make the compounds very promising for applications in the liquid nitrogen range and even at higher temperatures. Unfortunately, bulk samples of this material can only be synthesized under high pressure [11–14]. The sample size is typically a few millimeters, which limits its large-scale application.

Thin-film deposition methods are sometimes effective in stabilising transient high-pressure phases; Table 1 shows the progress of (Cu,C)-1234 films. The preparation of (Cu,C)-1234 films by pulsed laser deposition (PLD) [15–21], sputtering [22], and molecular beam epitaxy (MBE) [23,24] has been extensively studied. Compared to other methods, PLD has the advantage that there is no segregation of the film components and that the growth conditions can be controlled very well during film growth. Currently, (Cu,C)-1234 films can only be prepared by epitaxy on $LaAlO_3$(*l*00) single crystal substrates by PLD and on $NdGaO_3$(*l*00) and $SrLaGaO_4$(00*l*) single crystal substrates by MBE. In order to produce high-quality (Cu,C)-1234 films, it is important to investigate the effects of different substrates and different deposition temperatures on the epitaxial growth, structural, and electrical properties of (Cu,C)-1234 films. Since the $SrLaAlO_4$ substrate is characterized by high-temperature resistance, it does not react with (Cu,C)-1234 at higher temperatures (475 °C–900 °C). Meanwhile, the lattice constants and thermal expansion coefficients of $SrLaAlO_4$ substrates are similar to those of (Cu,C)-1234 films, which is favorable for the growth of (Cu,C)-1234 films. This paper focuses on trying out the possibility of growing (Cu,C)-1234 films on $SrLaAlO_4$ substrates.

Table 1. The progress of (Cu,C)-1234 films.

Method	Film Types	Substrate	T_c^{onset} (K)	T_{c0} (K)	References
PLD	$Ba_2Ca_3Cu_4CO_3O_8$	$LaAlO_3$(*l*00)	80–100	58	[15–17]
	$Ba_2Ca_3Cu_4O_8CO_3$		110	75	[18]
	(Cu,C)-1234		115	78	[19]
	(Cu,C)-1234		118	96	[20,21]
Sputtering	$Ba_2CuO_2(CO_3)$	$SrTiO_3$(00*l*)	40-50	4.2	[22]
MBE	(Cu,C)-1234	$NdGaO_3$(*l*00)	105	55	[23,24]
		$SrLaGaO_4$(00*l*)	90	15	

In this work, we employed the PLD method to fabricate thin films of (Cu,C)-1234 on $SrLaAlO_4$ substrates by changing the deposition temperature. The purity of the phase, the quality of growth, and superconductivity were investigated using various experimental techniques, including XRD, scanning electron microscopy (SEM)m and a physical property measurement system. Our findings reveal an optimal film temperature of around 680 °C, and the structural, morphological, and physical characterization of these films is also reported.

2. Materials and Methods

2.1. Preparation of the (Cu,C)-1234 Target

The ceramic $Ba_2Ca_3Cu_{4.2}O_y$ target was produced using a conventional solid-state sintering process. Reagent grade oxide/carbonate powders, namely $BaCO_3$, $CaCO_3$, and CuO, were

selected as raw materials and weighed according to their stoichiometric ratios. These powders were then added to a 500 mL nylon vessel containing ethanol as solvent and various zirconia beads. The mixed slurry was then ground in a planetary ball mill at 400 rpm for 14 h. After drying at 100 °C for 4 h, the mixture was uniaxially pressed into discs (with a diameter of about 28 mm) and then calcined in a muffle furnace at 860 °C for 12 h. The calcined discs were then ground and ball-milled for 14 h in the same way as before calcination. The slurry was then dried, sieved, and pressed into a disc (with a diameter of about 28 mm and a thickness of about 5 mm) under a pressure of 66 MPa using a dry press. Finally, the $Ba_2Ca_3Cu_{4.2}O_y$ target was obtained after sintering at 880 °C for 12 h.

2.2. Deposition of the (Cu,C)-1234 Films

With a KrF excimer laser (λ = 248 nm, COHERENT Compex Pro 205F, Coherent Inc., Santa Clara, CA, USA), the (Cu,C)-1234 films were deposited using the PLD method on $SrLaAlO_4(00l)$ substrates with a target-to-sample distance of around 6.5 cm. The laser energy and frequency were 320 mJ $pulse^{-1}$ and 5 Hz. The deposition time was 20 min. The deposition temperature was 650 °C to 720 °C; 30 cm^3 min^{-1} of O_2 and 20 cm^3 min^{-1} of CO_2 were separately introduced into the deposition chamber, and the chamber pressure was controlled at 20 Pa during the deposition process. After deposition, the films were slowly cooled down in an oxygen atmosphere of 70 KPa to 500 °C, held at this temperature for 60 min, and then cooled down to room temperature at a rate of 8 °C min^{-1}.

2.3. Characterization Techniques

The crystallographic structure and preferential orientation of the (Cu,C)-1234 films were measured by X-ray diffraction (XRD, SmartLab, Rigaku, Tokyo, Japan) with a Cu Kα source (λ = 1.541 Å) and diffraction angles (2θ) from 15° to 60°. The scanning electron microscope images of the (Cu,C)-1234 films were examined using a field emission scanning electron microscope (FE-SEM, HITACHI-SU5000, Tokyo, Japan) in the top view. Electrical resistance measurements were performed with the Quantum Design Instrument PPMS-9 T (Physical Property Measurement System 9 T) using the standard four-probe method.

3. Results and Discussion

Figure 1a shows the XRD θ-2θ scan of the (Cu,C)-1234 films deposited on $SrLaAlO_4$ substrates at different substrate temperatures. The peaks corresponding to the (100) orientation of the (Cu,C)-1234 films were detected within the substrate temperature range of 660 °C to 710 °C. Whereas the peaks related to the (200) orientation were observed between 650 °C and 710 °C. Figure 1b shows the temperature dependence of the (001) and (002) peak intensities of the (Cu,C)-1234 films. The peak intensity was temperature-dependent. Figure 1 illustrates that the intensities of the (001) and (002) peaks of the (Cu,C)-1234 films decrease notably with rising temperature when the substrate temperature exceeds 690 °C, finally disappearing at 720 °C. Conversely, the intensities of the peaks of $BaCuO_2$ and Ca_2CuO_3 show a significant increase with temperature. The peak intensities of (001) and (002) of the (Cu,C)-1234 films increase significantly with temperature when the substrate temperature is below 680 °C. Conversely, the peak intensities of $BaCuO_2$ and Ca_2CuO_3 decrease significantly with increasing temperature. The XRD spectra show weak Ca_2CuO_3 peaks and $BaCuO_2$ peaks, which could be unreacted particles in the $Ba_2Ca_3Cu_{4.2}O_y$ target. Fitting the θ-2θ diffractograms revealed that the a-axis lattice constant of the (Cu,C)-1234 thin film is 3.859 Å, which is very close to the bulk a-axis lattice constant of 3.860 Å [12–14]. In addition, the (Cu,C)-1234 thin film had a black surface with an almost metallic lustre.

Analyzing the relationship between the peak intensities of the XRD peaks (100) and (200) of the (Cu,C)-1234 films and the substrate temperature, it can be concluded that the substrate temperature has a great influence on the crystalline quality of the films. This is because at a low substrate temperature, the atoms adsorbed on the substrate surface have a lower energy and a poorer ability to migrate on the substrate surface and are covered by other atoms before they reach the ideal nucleation position [25,26]. This leads to more defects in the film and poorer film orientation and crystal quality. As the substrate temperature increases, the residence time of the deposited atoms on the substrate decreases, but the diffusion rate and the total accessible area of diffusing atoms increase. This favors the nucleation and growth of the film and reduces the defects during film growth, which improves the crystal quality of the (Cu,C)-1234 films.

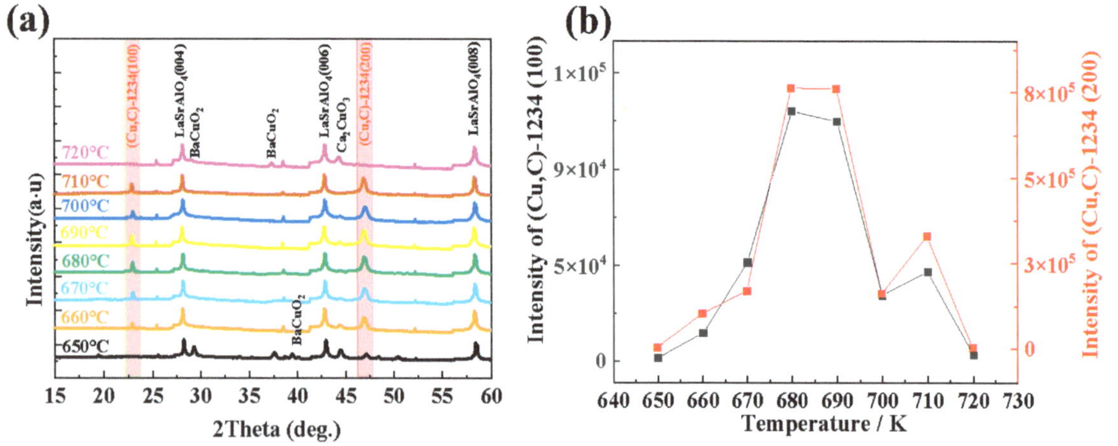

Figure 1. (**a**) The XRD θ-2θ scan of the (Cu,C)-1234 films deposited on $SrLaAlO_4$ substrates at different substrate temperatures; (**b**) the temperature dependence of the (001) and (002) peak intensities of the (Cu,C)-1234 films.

The phi-scans and the omega-scans make it possible to determine the quality of the (Cu,C)-1234 thin film. Figure 2 shows the omega and phi scan spectra of the (Cu,C)-1234 films deposited on $SrLaAlO_4$ films. The φ-scan of the (Cu,C)-1234 (104) film in Figure 2a shows four peaks at 90° intervals, indicating that the film is orientated and grew. As shown in Figures 1 and 2, the (Cu,C)-1234 thin film is purely a-axially orientated and exhibits good crystalline quality. The in-plane and out-of-plane half-height widths of 0.39° and 2.09°, respectively, indicate that the (Cu,C)-1234 films are epitaxially grown on $SrLaAlO_4$ substrates.

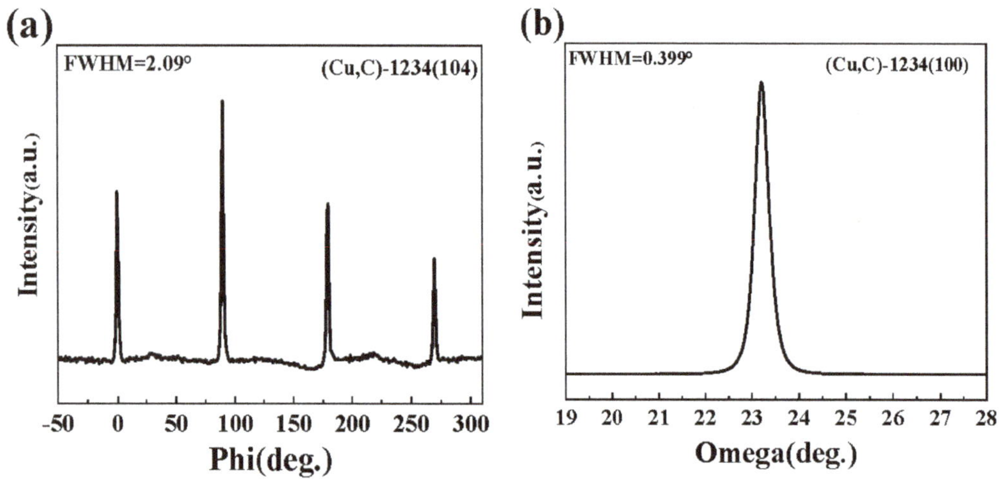

Figure 2. (Cu,C)-1234 thin film with a substrate temperature of 680 °C: (**a**) (104) phi scan curves in-plane; (**b**) (100) omega scan curves out-of-plane.

The critical temperature and the critical current density of superconducting materials are strongly dependent on the surface morphology of the materials. In order to clarify the morphology of the (Cu,C)-1234 films, the surface morphology was analyzed by SEM. Figure 3 shows SEM images of the (Cu,C)-1234 films prepared at different substrate temperatures. Figure S1 shows SEM images of the $SrLaAlO_4$ substrate. As can be seen in Figure 3, the surface of the (Cu,C)-1234 films is free of cracks at different temperatures. At substrate temperatures of 650 °C, 660 °C, and 670 °C, there are only a few rice-like (Cu,C)-1234 grains on the surface of the film; as the temperature increases, the rice-like (Cu,C)-1234 grains gradually increase and show a continuous shape, and the (Cu,C)-1234 grains are perpendicular to each other and show a twin structure with 90°. The 90°-orientated twin structure is a typical feature of this material [23,24]. However, when the substrate temperature reaches 720 °C, the surface of the film becomes wrinkled and the (Cu,C)-1234 grains disappear completely. The combination of the above experimental results shows that 680 °C is the optimal substrate temperature for the epitaxial growth of (Cu,C)-1234 films.

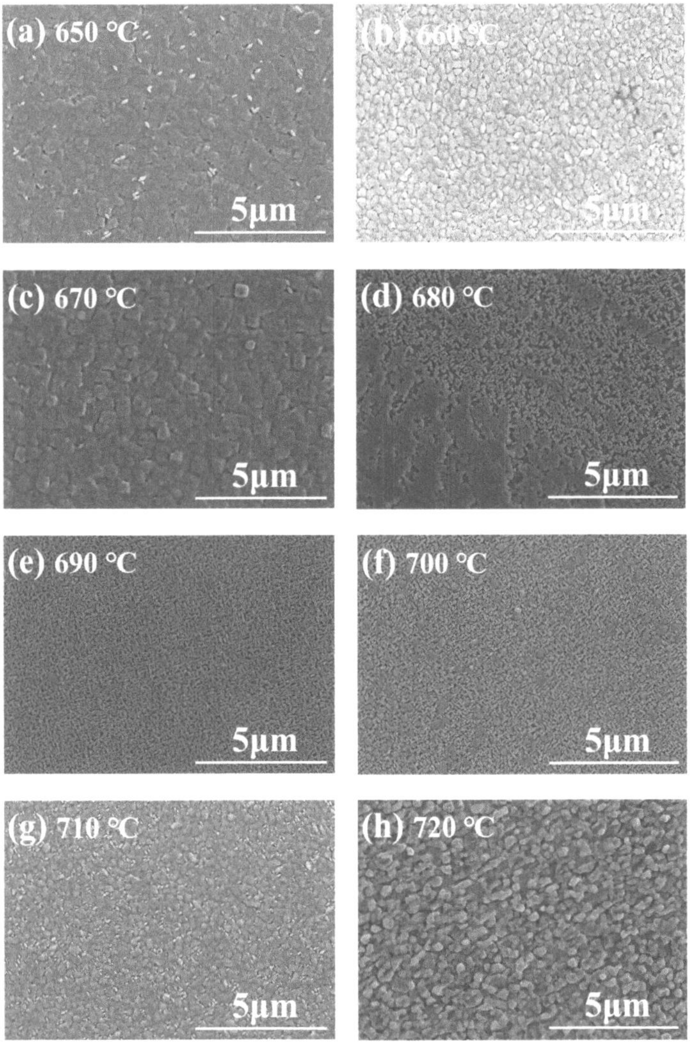

Figure 3. SEM images of the (Cu,C)-1234 films grown on $SrLaAlO_4(00l)$ substrates at different substrate temperatures.

Figure 4 shows the temperature dependence of resistivity for the (Cu,C)-1234 film with a deposition temperature of 680 °C under zero magnetic field, with the red line in the figure showing *dR/dT*. Figure 4 shows that the samples exhibit metallic properties until the superconducting transition occurs. In this work, T_c^{onset} was determined using the point of deviation from the linear range of the metal as a criterion and T_c was determined using the point of $dR/dT = 0$ as a criterion; T_c^{onset} was about 116 K, T_c was about 60 K, and T_{c0} was about 53 K. The T_{c0} of the epitaxially grown (Cu,C)-1234 films were low and not as good as those of the (Cu,C)-1234 blocks prepared by the high-pressure method. One of the reasons for this could be that the process of producing (Cu,C)-1234 films on $SrLaAlO_4$ substrates is not optimal and there is still a certain amount of non-superconducting phases, such as Ca_2CuO_3 and $BaCuO_2$, which need to be further explored and optimized to improve the superconducting properties of the (Cu,C)-1234 films, including the deposition temperature, CO_2 flow rate, and composition of the target materials and other parameters. The second reason could be that the fabricated (Cu,C)-1234 films are *a*-axis oriented, i.e., the *a*-axis is perpendicular to the film surface, and there is a mutually perpendicular 90° twinning structure within the surface. Later attempts can be made to grow (Cu,C)-1234 with *c*-axis orientation to overcome this twinning problem. The third reason could be the non-uniform distribution of oxygen content in the (Cu,C)-1234 films, which can be solved by post-annealing the (Cu,C)-1234 films under an oxygen atmosphere.

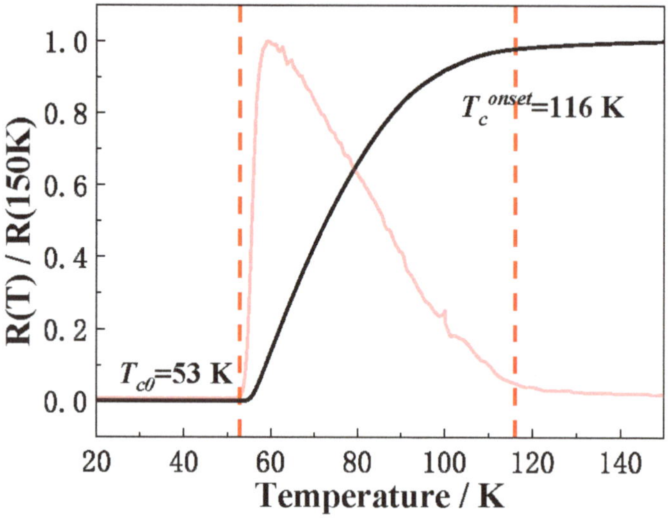

Figure 4. Temperature dependence of the resistance for the (Cu,C)-1234 film with a deposition temperature of 680 °C at a magnetic field of zero; the black line shows the resistance-temperature curve; the solid pink line shows the differential resistance as a function of temperature; the red dotted lines represent the lines at temperatures of 116 K and 53 K respectively.

Figure 5a shows the resistance versus temperature curves of the (Cu,C)-1234 films at different external fields, and the inset shows dR/dt at different magnetic fields. Figure 5a shows that the (Cu,C)-1234 films have a T_{c0} value of about 53 K and a ΔT_c value of about 36 K at zero field. The high ΔT_c value indicates that the superconducting grains are poorly bonded to each other and the quality of the film needs to be optimized. As shown in Figure 5a, the T_c value of the samples gradually decreases with the increase of the magnetic field, and the ΔT_c value gradually increases with the increase of the magnetic field. The upper critical field H_{c2} and the H_{irr} of the (Cu,C)-1234 films were determined using 90% R_n and 0.1% R_n as the criterion, and the results are shown in Figure 5b. As can be seen in Figure 5b, it is also difficult to observe a significant change in the T_c^{onset} when the magnetic field is increased to 9 T, indicating that there is a large upper critical field $H_{c2}(0)$ in the zero-temperature limit. We have not yet observed the H_{irr} at the temperature of liquid nitrogen, but the H_{irr} reached 9 T at 37 K. The main reason for the low H_{irr} of this film is the presence of a certain amount of heterogeneous phases, the 90° twin structure, etc., in the prepared film, which can significantly affect the superconducting properties, so the quality of the film needs to be further improved.

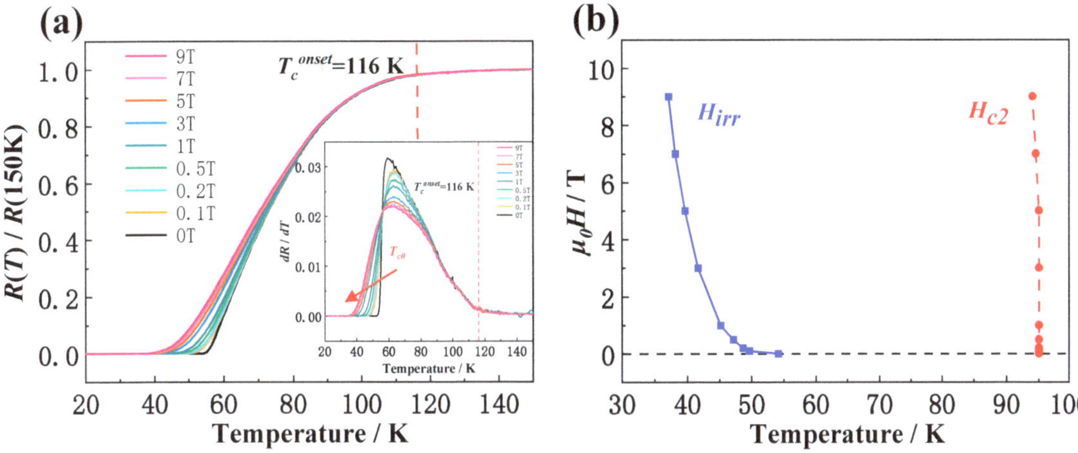

Figure 5. (**a**) Temperature dependence of the resistivity at zero field and under magnetic fields of H//*a* for the (Cu,C)-1234 films; the inset shows *dR/dT* at different magnetic fields. (**b**) Upper critical field and H_{irr} determined by the criterion of 90% and 0.1% of the normal state value for the (Cu,C)-1234 thin film. Red circles show H_{c2} and blue squares show H_{irr}.

4. Conclusions

Epitaxial growth of (Cu,C)-1234 films on $SrLaAlO_4(00l)$ single crystal substrates was successfully achieved for the first time through PLD. This study highlights the significant impact of deposition temperature on the quality of growth and surface morphology of (Cu,C)-1234 films, identifying the optimal deposition temperature as 680 °C. The peak intensity of (Cu,C)-1234(*l*00) exhibits a non-linear relationship with deposition temperature, showing an initial increase followed by a decrease, while the number of grains follows a similar trend. The in-plane and out-of-plane textures of the (Cu,C)-1234 films grown at the optimal temperature are measured at 2.09° and 0.39°, respectively, with a T_c^{onset} of around 116 K, a T_{c0} of approximately 53 K, and a H_{irr} of 9 T at 37 K. These findings provide valuable insights for further research into the growth mechanism of (Cu,C)-1234 films, with potential implications for the large-scale production and practical utilization of (Cu,C)-1234 superconductor materials.

An investigation concerning the growth of (Cu,C)-1234 films on different substrates is in progress. With further research on the phase formation mechanism and process conditions of (Cu,C)-1234 films prepared by PLD, the superconductivity of the film can be significantly enhanced. There is potential for the epitaxial growth of *c*-axis oriented (Cu,C)-1234 films in the near future, maximizing the performance of (Cu,C)-1234 superconducting materials. This could lead to the development of high-performance, low-cost, and reliable high-temperature superconducting strips by combining this technology with second-generation high-temperature superconducting strip technology. The future prospects for these new (Cu,C)-1234 high-temperature superconducting strips are very promising and could greatly contribute to the advancement of superconducting materials.

Supplementary Materials: The following supporting information can be downloaded at: https://www.mdpi.com/article/10.3390/cryst14060514/s1, Figure S1: SEM images of $SrLaAlO_4$ substrates.

Author Contributions: Z.L. and C.C. conceived and executed the project. Y.L. performed the thin film deposition and XRD, SEM, and R-T measurements; analyzed the data; and wrote the manuscript with the help of P.Z. and J.H. Z.L. analyzed the data and reviewed the manuscript. C.C. provided funding support for the project and reviewed the manuscript. All authors have read and agreed to the published version of the manuscript.

Funding: This work was supported by the National Natural Science Foundation of China (Grant Nos.52172271, 12374378, 52307026), the National Key R&D Program of China (Grant No. 2022YFE03150200), the Shanghai Science and Technology Innovation Program (Grant No. 22511100200, 23511101600), and the Strategic Priority Research Program of the Chinese Academy of Sciences (Grant No. XDB25000000).

Data Availability Statement: The raw data supporting the conclusions of this article will be made available by the authors on request.

Acknowledgments: We thank Haihu Wen for helpful and stimulating discussions.

Conflicts of Interest: The authors declare no conflicts of interest.

References

1. Bondarenko, S.I.; Koverya, V.P.; Krevsun, A.V.; Link, S.I. High-temperature superconductors of the family (RE)$Ba_2Cu_3O_{7-\delta}$ and their application. *Low Temp. Phys.* **2017**, *43*, 1125. [CrossRef]
2. Cai, C.; Chi, C.; Li, M.; Liu, Z.; Lu, Y.; Guo, Y.; Bai, C.; Lu, Q.; Dou, W. Advance and challenge of secondary-generation high- temperature superconducting tapes for high field applications. *Sci. Bull.* **2019**, *64*, 827. [CrossRef]
3. Wang, K.; Hou, Q.; Pal, A.; Wu, H.; Si, J.; Chen, J.; Yu, S.; Chen, Y.; Lv, W.; Ge, J.; et al. Structural and Physical Properties of High-Entropy $REBa_2Cu_3O_{7-\delta}$ Oxide Superconductors. *J. Supercond. Novel Magn.* **2021**, *34*, 1379. [CrossRef]
4. Cai, C.; Li, Z.; Lu, Y. Evolvement and Prospect of Practical Superconducting Materials. *Mater. China* **2011**, *30*, 1.
5. Li, Q.; Suenaga, M.; Kaneko, T.; Sato, K.; Simmon, C. Collapse of irreversible field of superconducting $Bi_2Sr_2Ca_2Cu_3O_{10+\delta}$/Ag tapes with columnar defects. *Appl. Phys. Lett.* **1997**, *71*, 1561. [CrossRef]
6. Clayton, N.; Musolino, N.; Giannini, E.; Garnier, V.; Flükiger, R. Growth and superconducting properties of $Bi_2Sr_2Ca_2Cu_3O_{10}$ single crystals. *Supercond. Sci. Technol.* **2004**, *17*, S563. [CrossRef]
7. Liu, J.; Cheng, J.; Wang, Q. Analysis and enhancement of cooling system of high-speed permanent magnet motor based on computational fluid dynamics. *Adv. Technol. Electral. Eng. Energy* **2017**, *36*, 1.
8. Dadras, S.; Falahati, S.; Dehghani, S. Effects of graphene oxide doping on the structural and superconducting properties of $YBa_2Cu_3O_{7-\delta}$. *Phys. C* **2018**, *548*, 65. [CrossRef]
9. Palau, A.; Valles, F.; Rouco, V.; Coll, M.; Li, Z.; Pop, C.; Mundet, B.; Gazquez, J.; Guzman, R.; Gutierrez, J.; et al. Disentangling vortex pinning landscape in chemical solution deposited superconducting $YBa_2Cu_3O_{7-x}$ films and nanocomposites. *Sci. Technol.* **2018**, *31*, 034004.
10. Johnson, C.L.; Bording, J.K.; Zhu, Y. Structural inhomogeneity and twinning in $YBa_2Cu_3O_{7-\delta}$ superconductors: High-resolution transmission electron microscopy measurements. *Phys. Rev. B* **2008**, *78*, 14517. [CrossRef]
11. Shimakawa, Y.; Jorgensen, J.D.; Hinks, D.G.; Shaked, H.; Hitterman, R.L.; Izumi, F.; Kawashima, T.; Takayama-Muromachi, E.; Kamiyama, T. Crystal structure of (Cu,C)$Ba_2Ca_3Cu_4O_{11+\delta}$($T_c$ = 117 K) by neutron-powder-diffraction analysis. *Phys. Rev. B* **1994**, *50*, 16008. [CrossRef] [PubMed]
12. Zhang, Y.; Liu, W.; Zhu, X.; Zhao, H.; Hu, Z.; He, C.; Wen, H. Unprecedented high irreversibility line in the nontoxic cuprate superconductor (Cu,C)$Ba_2Ca_3Cu_4O_{11+\delta}$. *Sci. Adv.* **2018**, *4*, eaau0192. [CrossRef] [PubMed]
13. He, C.; Ming, X.; Si, J.; Zhu, X.; Wang, J.; Wen, H. Characterization of the (Cu,C)Ba2Ca3Cu4$O_{11+\delta}$ single crystals grown under high pressure. *Supercond. Sci. Technol.* **2021**, *35*, 025004. [CrossRef]
14. He, C.; Ming, X.; Lin, R.; Fan, X.; Song, D.; Ge, B.; Wen, H. Key factor for low anisotropy and high irreversibility field in (Cu,C)$Ba_2Ca_3Cu_4O_{11+\delta}$. *Mater. Today Phys.* **2022**, *29*, 100913. [CrossRef]
15. Allen, J.L.; Mercey, B.; Prellier, W.; Hamet, J.F.; Hervieu, M.; Raveau, B. The first superconducting thin film oxycarbonates coherent intergrowth of several members of a new structural series $(CaCuO_2)_m(Ba_2CuO_2CO_3)_n$. *Phys. C* **1995**, *241*, 158. [CrossRef]
16. Prellier, W.; Allen, J.L.; Prouteau, C.; Simon, C.; Mercey, B. Irreversibility line and critical current densities of an oxycarbonate thin film $Ba_2Ca_3Cu_4CO_3O_8$. *Supercond. Sci. Technol.* **1995**, *8*, 361. [CrossRef]
17. Hervieu, M.; Mercey, B.; Prellier, W.; Allen, J.L.; Hamet, J.F.; Raveau, B. Microstructure of superconducting copper oxycarbonate thin films of the Ba–Ca–Cu–C–O system. *J. Mater. Chem.* **1996**, *6*, 165. [CrossRef]
18. Prellier, W.; Tebano, A.; Allen, J.L.; Mercey, B.; Hamet, J.F.; Hervieu, M.; Raveau, B. High T_c superconducting oxycarbonate thin films grown by laser ablation: Promising new materials. *Phys. C* **1997**, *282*, 647. [CrossRef]

19. Calestani, G.; Migliori, A.; Spreitzer, U.; Hauser, S.; Fuchs, M.; Barowski, H.; Schauer, T.; Assmann, W.; Range, K.J.; Varlashkin, A.; et al. Ba-Ca–Cu oxycarbonate thin films, prepared by pulsed laser deposition: Structure, growth mechanism and superconducting properties. *Phys. C* **1999**, *312*, 225. [CrossRef]
20. Duan, T.; Hao, J.; Chu, H.; Li, B.; Dai, Y.; Wen, H. Existence of carbonate clusters and its relationship with critical temperature in superconducting (Cu,C)$Ba_2Ca_3Cu_4O_y$ films. *Phys. C* **2020**, *573*, 1353646. [CrossRef]
21. Duan, T.; Hao, J.; Chu, H.; Wen, H. Preparation and superconducting properties of the (Cu,C)$Ba_2Ca_3Cu_4O_{11+y}$ films with zero-resistance transition temperature of 96 K. *Supercond. Sci. Technol.* **2020**, *33*, 025009. [CrossRef]
22. Adachi, H.; Sakai, M.; Satoh, T.; Setsune, K. Epitaxial Growth and Properties of (CO_3) Stabilised $BaCuO_2$ Superconducting Thin Films. In Proceedings of the 8th International Symposium on Superconductivity VIII. Advances in Superconductivity VIII, Hamamatsu, Japan, 30 October–2 November 1996; p. 955.
23. Shibata, H.; Karimoto, S.; Tsukada, A.; Makimoto, T. Growth of (Cu,C)$Ba_2Ca_{(n-1)}Cu_nO_y$ thin films by molecular-beam epitaxy. *Phys. C* **2006**, *445*, 862. [CrossRef]
24. Shibata, H.; Karimoto, S.; Tsukada, A.; Makimoto, T. Thin film growth of (Cu,C)$Ba_2Ca_{(n-1)}Cu_nO_y$(n=1–4) superconductor by molecular beam epitaxy. *J. Crys. Growth* **2007**, *301*, 684–686. [CrossRef]
25. Shi, W.; Shi, J.; Sun, J.; Yao, W.; Qi, Z. Influence of substrate temperature on the orientation of $YBa_2Cu_3O_7$ films. *Appl. Phys. Lett.* **1990**, *5*, 822. [CrossRef]
26. De, R.; Augustine, S.; Das, B.; Sikdar, M.K.; Ranjan, M.; Sahoo, P.K.; Haque, S.M.; Prathap, C.; Rao, K.D. Influence of in-situ substrate temperature on anisotropic behaviour of glancing angle grown nickel nanocolumns. *Appl. Phys. A* **2024**, *130*, 126. [CrossRef]

crystals

MDPI

Article

Ultrahigh Responsivity In_2O_3 UVA Photodetector through Modulation of Trimethylindium Flow Rate

Yifei Li [1,2], Tiwei Chen [2,3], Yongjian Ma [2,3], Yu Hu [2,3], Li Zhang [2], Xiaodong Zhang [2,3], Jinghang Yang [1], Lu Wang [1,2], Huanyu Zhang [2,3], Changling Yan [1,*], Zhongming Zeng [2,3] and Baoshun Zhang [2,3,*]

1 State Key Laboratory of High-Power Semiconductor Laser, School of Physics, Changchun University of Science and Technology, Changchun 130022, China; yfli2022@sinano.ac.cn (Y.L.); 2020200035@mails.cust.edu.cn (J.Y.); lwang2022@sinano.ac.cn (L.W.)

2 A Nanofabrication Facility, Suzhou Institute of Nano-Tech and Nano-Bionics, Chinese Academy of Sciences, Suzhou 215123, China; twchen2020@sinano.ac.cn (T.C.); yjma2019@sinano.ac.cn (Y.M.); yhu2022@sinano.ac.cn (Y.H.); lizhang2017@sinano.ac.cn (L.Z.); xdzhang2007@sinano.ac.cn (X.Z.); huanyuzhang@mail.ustc.edu.cn (H.Z.); zmzeng2012@sinano.ac.cn (Z.Z.)

3 School of Nano Technology and Nano Bionics, University of Science and Technology of China, Hefei 230026, China

* Correspondence: changling_yan@126.com (C.Y.); bszhang2006@sinano.ac.cn (B.Z.)

Abstract: Oxygen vacancies (V_o) can significantly degrade the electrical properties of indium oxide (In_2O_3) thin films, thus limiting their application in the field of ultraviolet detection. In this work, the V_o is effectively suppressed by adjusting the Trimethylindium (TMIn) flow rate (f_{TMIn}). In addition, with the reduction of the f_{TMIn}, the background carrier concentration and the roughness of the film decrease gradually. And a smooth In_2O_3 thin film with roughness of 0.44 nm is obtained when the f_{TMIn} is 5 sccm. The MSM photodetectors (PDs) are constructed based on In_2O_3 thin films with different f_{TMIn} to investigate the opto-electric characteristics of the films. The dark current of the PDs is significantly reduced by five orders from 100 mA to 0.28 μA with the reduction of the f_{TMIn} from 50 sccm to 5 sccm. In addition, the photo response capacity of PDs is dramatically enhanced. The photo-to-dark current ratio (PDCR) increases from 0 to 2589. Finally, the PD with the f_{TMIn} of 5 sccm possesses a record-high responsivity of 2.53×10^3 AW^{-1}, a high detectivity of 5.43×10^7 Jones and a high EQE of 9383 × 100%. Our work provides an important reference for the fabrication of high-sensitivity UV PDs.

Keywords: In_2O_3 thin films; oxygen vacancy; photodetector; TMIn flow rate

Citation: Li, Y.; Chen, T.; Ma, Y.; Hu, Y.; Zhang, L.; Zhang, X.; Yang, J.; Wang, L.; Zhang, H.; Yan, C.; et al. Ultrahigh Responsivity In_2O_3 UVA Photodetector through Modulation of Trimethylindium Flow Rate. *Crystals* **2024**, *14*, 494. https://doi.org/10.3390/cryst14060494

Academic Editor: Anelia Kakanakova

Received: 20 April 2024
Revised: 20 May 2024
Accepted: 21 May 2024
Published: 24 May 2024

1. Introduction

Ultraviolet-A (UVA) photodetectors (PDs) have attracted wide attention in both civilian and military applications, such as flame detection, ultraviolet cameras, and missile early warning systems [1–9]. Indium compounds such as In_2Se_3, In_2S_3 and In_2O_3 are widely reported for efficient photoelectric detection [10–18]. Because metal oxides can be used as interface layers in metal/semiconductor contacts to change the optoelectronic properties of heterojunctions, they are widely used in optoelectronic devices [19]. Therefore, In_2O_3 has been considered a promising candidate for UVA PD fabrication among indium compounds, due to its suitable bandgap (3.7 eV), high mobility and stable chemical and optical properties [20–22]. Currently, UVA PDs based on In_2O_3 thin films mainly depends on the photovoltaic effect, because it can realize the efficient separation and collection of photogenerated carriers through the built-in electric field and reduce the dark current of the device [23–25]. The device structure based on photovoltaic effect mainly includes heterojunctions, transistors and metal–semiconductor–metal (MSM) structures. Among these, MSM PDs offer competitive advantages such as simple structure, ease of fabrication and convenient circuit integration [26].

In_2O_3 MSM PDs are typically integrated with circuits through flip-chip bonding, enabling their application in UV imaging via back illumination. Therefore, low-cost and UVA-transparent sapphire substrates are widely used for In_2O_3 thin film detectors. In addition, the epitaxy of high-crystal-quality In_2O_3 thin films has been reported through HVPE, MBE and MOCVD. Due to its ability to produce high-uniform and reproducible thin films, MOCVD offers advantages such as rapid growth rates that enhance production efficiency and reduce costs for the commercialization of high-quality In_2O_3 thin films and their detectors. However, defects such as oxygen vacancies (V_o) in the In_2O_3 thin film can deteriorate the performance of the PDs, which significantly limits their application [27]. It has been reported that the V_o has a significant effect on the metal/semiconductor interface and leads to a decrease in the Schottky barrier, resulting in a significant increase in the dark current of the detector [28]. In addition, the ionized oxygen vacancy can capture photogenerated electrons, and the release process is very slow after the light source is turned off, resulting in a persistent photoconductive effect and degradation of the response speed [29]. Ge et al. proposed hydrogen (H) doping in In_2O_3 through magnetron sputtering and proved H can occupy the V_o [30]. With the H doping concentration of 1%, the thin film showed a high mobility of 115.3 $cm^2\ V^{-1}s^{-1}$. However, H exhibits low stability in In_2O_3 thin films and tends to escape at high temperatures.

In this work, we significantly suppress the defects of V_o in In_2O_3 thin films by adjusting the Trimethylindium (TMIn) flow rate (f_{TMIn}) during film growth. With a reduction in the f_{TMIn}, not only the concentration of V_o is suppressed, but also the performance of the In_2O_3 PDs is improved. Consequently, the dark current of the PDs based on the In_2O_3 thin film decreased by five magnitudes. In addition, the UV PD exhibits a high responsivity (2.53×10^3 AW^{-1}) and a high detectivity (5.43×10^7 Jones). This work provides an important reference for the fabrication of high-performance UVA PDs based on In_2O_3 thin film.

2. Materials and Methods

The unintentionally doped (UID) In_2O_3 thin films are grown on sapphire substrates using MOCVD. The growth is performed at 600 °C for 120 min. TMIn is used as the In precursor and the flow rate of O_2 (99.999%) is fixed at 1000 sccm. High-purity nitrogen (N_2) serves as the carrier gas. To construct a UVA PD, the Pt/Au (50/100 nm) bilayer metal is deposited on the In_2O_3 thin film by electron beam evaporation. Finally, the electrodes are patterned into intercalated fingers with a length of 480 µm, a width of 10 µm and a spacing of 10 µm. The electrodes contain 12 pairs of fingers, and the area of the active region is 1.3×10^{-3} cm^2.

The thickness of the In_2O_3 thin films is measured to be approximately 50 nm using SEM as shown in Figure S1 in Supplementary Materials. The crystal structure and crystallinity of In_2O_3 thin films are characterized by X-ray diffraction (XRD) with a test voltage of 40 kV, a current of 40 mA and a step size of 0.02°/s The chemical compositions of the In_2O_3 thin films are analyzed by X-ray photoelectron spectroscopy (XPS). An XPS instrument with a base pressure better than 4.4×10^{-8} Pa is used to acquire core-level XPS spectra from the sample. Monochromatic Al Kα radiation source (hv = 1486.6 eV) is employed. The size of analyzed area is 100 × 100 µm and the optoelectronic detection angle was 45°. Argon ion etching is not carried out to avoid the influence of etching damage on the atomic ratio [31]. The surface morphologies are obtained by atomic force microscopy (AFM) with am amplitude of 200~300 mV and a scanning speed of 1 Hz. The carrier concentration and mobility are tested by Hall measurements. The optoelectronic performance of the devices is measured via a probe station equipped with Keysight B1505A (Colorado Springs, United States), and a commercial monochromatic tunable light sources system (Omno150300, Abuja, NigeriaNBET), including Xenon lamp, Monochromator, chopper and optical fibers.

3. Results

The In_2O_3 thin films grown at the f_{TMIn} of 50, 25, 10 and 5 sccm are labeled as S_1, S_2, S_3 and S_4, respectively. In order to study the effect of f_{TMIn} on the crystal structure and crystallinity of the In_2O_3 thin films, XRD measurements are carried out. As shown in Figure 1a, the diffraction peak observed near 41.68° corresponds to the (0006) plane of the sapphire substrate. In addition, two diffraction peaks located at 30.65° and 64° are observed, which correspond to the (222) and (444) planes of In_2O_3. The results indicate that the heteroepitaxial In_2O_3 thin films grown on c-plane sapphire have a preferred orientation along the (222) direction. The coherent volume of specific diffraction peaks in thin films is related to the size of crystallite [32]. According to Scherrer's relation, the size of the crystallites in S_1 to S_4 can be calculated using the following equation:

$$\text{Crystallite size (D)} = \frac{0.9\lambda}{\beta\cos\theta} \quad (1)$$

where λ = 1.54 Å is the wavelength of the Cu Kα line, θ is the Bragg's angle and β is the full width at half maximum (FWHM). Therefore, the D of S_1 to S_4 can be calculated to be 28.0 nm, 15.9 nm, 7.4 nm and 4.3 nm, respectively. In addition, the dislocation density can be expressed as

$$\text{Dislocation density } (\delta) = \frac{1}{D^2} \quad (2)$$

According to Equation (2), the δ of S_1 to S_4 can be estimated to be 1.27×10^{11} cm^{-2}, 3.95×10^{11} cm^{-2}, 1.83×10^{12} cm^{-2} and 5.53×10^{12} cm^{-2}, respectively. The size of the crystallites has a significant influence on the crystallite number (N_C). The N_C can be estimated using the following equation:

$$N_C = \frac{d}{D^3} \quad (3)$$

where d is the thickness of the thin film. Therefore, the N_C of S_1 to S_4 is 2.29×10^{11} cm^{-2}, 1.23×10^{12} cm^{-2}, 1.24×10^{13} cm^{-2} and 6.49×10^{13} cm^{-2}, respectively. With the reduction of the f_{TMIn}, the film crystallite size decreases significantly and the dislocation density increases. This is due to the fact that a large number of In atoms are deposited on the substrate surface with a high f_{TMIn}, which promotes the growth of crystallites through the diffusion and migration of the deposited atoms [33]. Therefore, larger crystallite size can be obtained at high f_{TMIn} in this study. The decrease of crystallite size leads to the increase of dislocation density, so the decrease of f_{TMIn} will reduce the crystal quality of thin films to some extent. As shown in Figure 1b, high-resolution XRD rocking curve of (222) plane is performed to identify the crystallinity of the In_2O_3 thin films. The FWHM for (222) plane of S_1 to S_4 are extracted to be 0.26°, 0.31°, 0.44° and 0.48°, respectively. The results indicate that the crystallinity of In_2O_3 thin films slightly decreased with the decrease of the f_{TMIn} due to the decrease of the grain size.

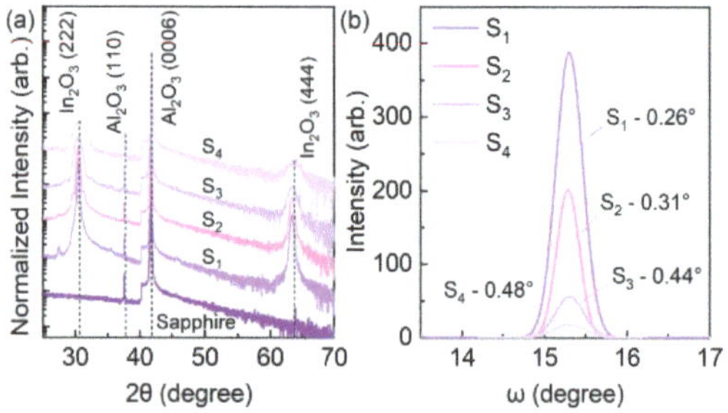

Figure 1. (**a**) XRD 2θ scan pattern and (**b**) XRD rocking curve of S_1 to S_4.

The AFM measurements with the scanning area of 5 × 5 μm^2 are carried out to analyze the surface morphology of the In_2O_3 thin films. As shown in Figure 2a, the large island structure can be observed on the surface of the film when the f_{TMIn} is 50 sccm. With the reduction of the f_{TMIn}, the size of the island structure gradually shrinks, and the density increases gradually. The results are consisted of that of XRD. Finally, the surface of the In_2O_3 thin films gradually becomes smooth and the roughness of S_1, S_2, S_3 and S_4 are 2.41 nm, 1.08 nm, 0.56 nm and 0.44 nm, respectively. It is worth noting that compared with the scattering of carriers through crystallite boundaries, the capture and release of carriers by point defects (e.g., V_o) have a more significant effect on the optical response of the PDs [34,35].

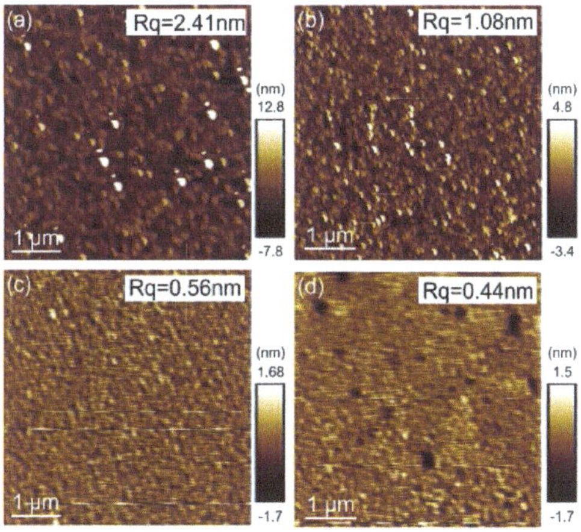

Figure 2. AFM images with a 5 × 5 μm^2 scan area of In_2O_3 thin film with the f_{TMIn} of (**a**) 50 sccm, (**b**) 25 sccm, (**c**) 10 sccm and (**d**) 5 sccm.

In order to study the effect of the f_{TMIn} on the defects in In_2O_3 thin films, the XPS measurements are carried out. Figure 3a shows the XPS spectra of the In_2O_3 thin films grown on sapphire substrates. The binding energy of C 1s peak plus the work function of In_2O_3 is a constant value of 289.58 eV, which can be used for the alignment of XPS spectra [36]. The work function of In_2O_3 is 5.0 eV, and the C 1s peak of In_2O_3 can be set at 284.58 eV [37,38]. As shown in Figure 3a, peaks of C 1s, O 1s, In 3d, In 3d3/2, In 3p, In 3p1/2 and In 4p signals are observed in the general scan range of 0 to 1200 eV. The result indicates that the chemical composition of the In_2O_3 thin films grown on sapphire mainly consists of In and O. The position of the O 1s peak is located at 529.58 eV. The O 1s spectrum can be deconvoluted into two components. The component located at 529.58 eV (O_I) corresponds to binding energy of lattice oxygen in the In_2O_3 thin film, while the component at 531.2 eV (O_{II}) corresponds to binding energy of defects (V_o) [39]. The V_o in the In_2O_3 thin film can lead to a deterioration in the photo response performance of the PDs and the oxygen-rich atmosphere is beneficial for suppression of the V_o. As shown in Figure 3b–e, the intensity ratios of $O_{II}/(O_I + O_{II})$ for S_1 to S_4 are 27.23%, 25.78%, 25.43% and 24.78%, respectively. The result indicates that the V_o is reduced with the reduction of the f_{TMIn}. The suppression of V_o in In_2O_3 thin films is due to the fact that with the decrease of the f_{TMIn}, the V_o sates are compensated by oxygen in oxygen-rich environment [40].

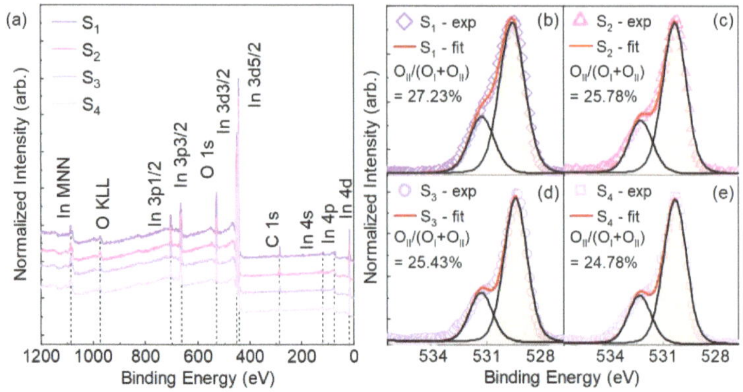

Figure 3. (**a**) Typical survey XPS spectra of S1 to S4. XPS O 1s core-level spectra of (**b**) S_1, (**c**) S_2, (**d**) S_3 and (**e**) S_4.

It is reported that the V_o in In_2O_3 thin films can provide electrons as shallow donors, thus UID-In_2O_3 thin films often contain very high background carrier concentration [41–43]. To evaluate the electrical properties of the In_2O_3 crystal film, Hall measurements are carried out and the measurement structure has been illustrated in Figure 4a. As shown in Figure 4c, the carrier concentration deceases significantly with the decrease of the f_{TMIn}, which is beneficial to suppress the dark current of the PDs. The carrier concentrations of S_1 to S_4 are 1.53×10^{19} cm^{-3}, 8.31×10^{18} cm^{-3}, 4.94×10^{18} cm^{-3} and 4.31×10^{16} cm^{-3}, respectively. It worth noting that the mobility also gradually decreases with the decrease of the f_{TMIn}. Since the dislocation density and number of crystallites of the In_2O_3 thin films increases with the decrease of the f_{TMIn}, the decrease of mobility is speculated to be attributed to the scattering of the grain boundary [30]. The optimization of crystallinity and mobility will be further studied in future work. In order to investigate the effect of the f_{TMIn} on the performance of the PDs, the MSM PDs based on S_1 to S_4 are constructed and tested. Figure 4b illustrate the Schematic diagram device structure. The PDs based on S_1 to S_4 are labeled as PD_1 to PD_4, respectively. To prevent the PDs from being damaged by too large a current, the current is limited to 100 mA in the test. Figure 4d shows the dark current (I_d) of the PDs and the inset shows the microscope image of the device structure. As shown in Figure 4d, the PD_1, PD_2 and PD_3 exhibit extremely high current due to the high carrier concentration and reach the current limits at 0.75 V, 0.95 V and 2.95 V, respectively. However, when the f_{TMIn} is reduced to 5 sccm, the I_d of the PDs decreases significantly to the microampere level. Compared with PD_1, the I_d of PD_4 is reduced by 4.5×10^5 times at 0.75 V. It is worth mentioning that with the decrease of the f_{TMIn}, the devices gradually show Schottky behavior. As shown in Figure S2a–c in Supplementary Materials, there is a linear relation between the I_d and voltage for PD_1 to PD_3. The results indicate that the Pt/Au bilayer metal forms Ohmic contact with In_2O_3 thin films when the f_{TMIn} is 50 sccm, 25 sccm and 10 sccm. And the resistance of PD_1 to PD_3 can be calculated to be 7.08 Ω, 8.68 Ω and 15.06 Ω. Due to the low resistance between Pt/Au bilayer metal and In_2O_3 thin films, the I_d of the PDs are extremely huge. However, the relationship between voltage and I_d deviates from linearity when the f_{TMIn} is 5 sccm as shown in Figure S2d in Supplementary Materials. This is because with the decrease of the f_{TMIn}, the concentration of V_o in In_2O_3 thin films decreases and Schottky behavior recovers gradually [35]. The I_d of the devices with Schottky behavior will be significantly reduced, and the photo response ability will be improved. The results are consistent with that of Hall and XPS measurements, which indicates that the reduction of the f_{TMIn} can effectively suppress the V_o thus reduce the I_d of the In_2O_3 PDs.

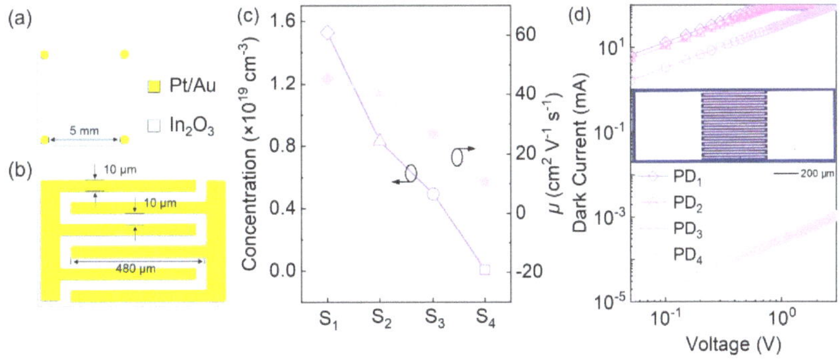

Figure 4. Schematic diagram of (**a**) Hall test structure and (**b**) In_2O_3 MSM PD. (**c**) Carrier concentration and mobility of S_1, S_2, S_3 and S_4. (**d**) The dark current of the PDs based on S_1, S_2, S_3 and S_4. (PDs based on S_1 to S_4 are labelled as PD_1 to PD_4).

Figure 5 shows the IV curves of PD_1 to PD_4 in the dark and under 335 nm illumination. Due to the large I_d, PD_1 and PD_2 exhibit almost no photo response as shown in Figure 5a,b. The photo-to-dark current ratio (PDCR) is an important figure-of-merit to evaluate the capacity of the photo response, which can be calculated by the following equation:

$$\mathrm{PDCR} = \frac{I_p - I_d}{I_d} \tag{4}$$

where I_p refers to photo current. Therefore, the PDCR of PD_1 and PD_2 can be calculated to be 0 in the test range. With the decrease of the carrier concentration, the photo response capacity of the PDs is obviously enhanced. The PDCR of PD_3 is calculated to be 0.94 at 1.5 V before the I_p reaches the limitation of the current. When the carrier concentration is 4.31×10^{16} cm^{-3} with the f_{TMIn} of 5 sccm, the PDCR of the PDs is dramatically improved. As shown in Figure 5d, PD_4 exhibits an obvious photo response with an extremely high PDCR of 2588 at 5 V. Furthermore, the responsivity (R) and external quantum efficiency (EQE) of the PD_4 are extracted using the following equation:

$$\mathrm{R} = \frac{I_p - I_d}{PS} \tag{5}$$

$$\mathrm{EQE} = \frac{Rhc}{e\lambda} \tag{6}$$

where P, S, h, c, e and λ refer to the intensity of the illumination (2000 μW/cm^2), the effective area, the Planck constant, the speed of the light, the electric charge and the wavelength, respectively. The R and EQE of PD_4 are 2.53×10^3 AW^{-1} and $9383 \times 100\%$, respectively. According to the results of XPS and Hall measurements, when the indium flow rate is 50 sccm, 25 sccm and 10 sccm, the concentration of V_o in the In_2O_3 film is higher, and the background carrier concentration is also higher due to the donor action of the V_o. Under this condition, the change of the device conductance caused by photogenerated minority carriers will not be obvious, so the device has almost no optical response. However, with the decrease of f_{TMIn}, the concentration of V_o decreases, the background carrier concentration of In_2O_3 thin film decreases significantly and the Schottky behavior is restored. The existence of the built-in electric field increases the resistance of the device, thus reducing the I_d of the device and improving the light response ability of the device. The above results indicate that reducing the background carrier concentration of the UID-In_2O_3 thin films by reducing the f_{TMIn} can effectively improve the photo response of the PDs.

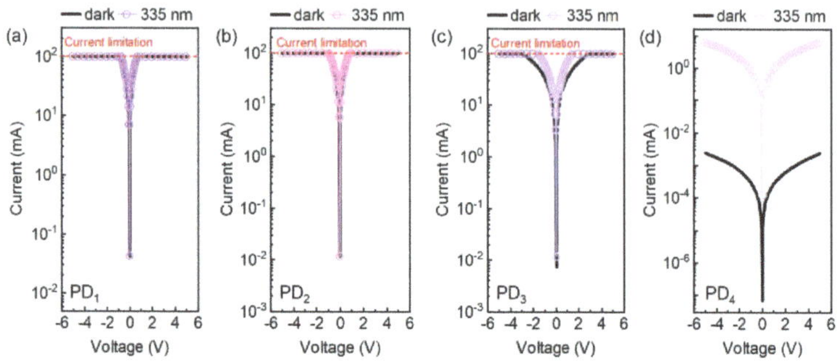

Figure 5. IV curves in the dark and under 335 nm illumination (2000 μW cm^{-2}) of (**a**) PD_1, (**b**) PD_2, (**c**) PD_3 and (**d**) PD_4.

In order to explain the effect of V_o on the performance of the PDs in detail, the energy band diagram of the In_2O_3 MSM PDs is shown in Figure 6. When the f_{TMIn} is high, the concentration of V_o in In_2O_3 thin films is also higher. On the one hand, the V_o as a donor state can increase the carrier concentration in In_2O_3 thin films, thus reducing the height of Schottky barrier. On the other hand, the V_o at the metal/In_2O_3 interface can cause Fermi pinning, which promotes the tunneling of electrons at the reverse biased Schottky junction, as the process 1 shown in Figure 6a [44,45]. What's more, some unoccupied states, such as ionized V_o states, can capture electrons, making it easier for electrons to tunnel through the barrier. This process is marked as process 2 and is depicted in the energy band structure of Figure 6a. It is worth mentioning that the V_o defects within the tunneling distance at and near the interface can promote the occurrence of tunneling. Finally, the electrons of the devices with high V_o concentration will mainly conform to the field emission or thermal field emission mechanism, rather than the thermal emission mechanism (process 3) [46,47]. This is why the PDs have Ohmic behavior rather than Schottky behavior and the I_d is extremely high. Under 335 nm illumination, the electrons in the In_2O_3 film will transition from the valence band to the conduction band, resulting in photogenerated electron-hole pairs, as shown in process 4 in Figure 6b. Photogenerated electron-hole pairs will be separated by external electric field and built-in electric field and collected by electrodes. However, in the process of drift to the electrode, the photogenerated electrons will be trapped on the surface of the Metal/In_2O_3 thin film, due to the existence of a large number of oxygen vacancies (process 5) [48]. When the light source is turned off, the trapped photogenerated carriers will be released. However, this process is very slow, resulting in the PDs cannot quickly return to the dark state, thus deteriorating the response speed of the PDs. Therefore, in order to obtain highly sensitive and high-speed In_2O_3 MSM UV PDs, the defects such as V_o in the film should be further suppressed, which will be further studied in the future.

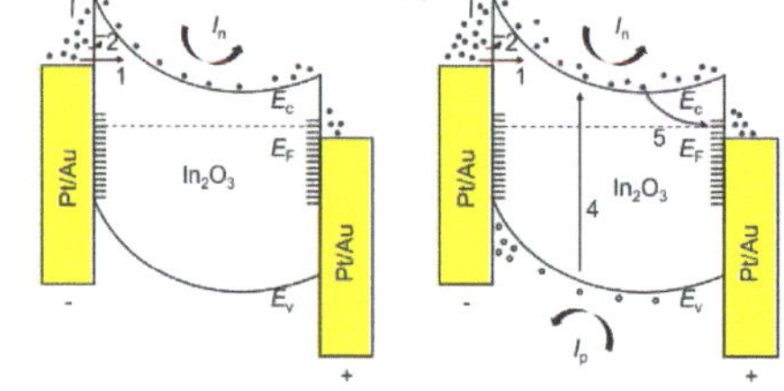

Figure 6. Schematic diagram of energy band structure of In_2O_3 MSM PD (**a**) under dark state and (**b**) under 335 nm illumination.

To obtain the specific detectivity (D*) of the PD_4, the noise characteristics are evaluated. By performing the Fourier transform of the dark current in the time domain, the noise power spectrum is obtained [49,50]. As shown in Figure 7a, the noise power density of PD4 is frequency dependent. The result indicates that the noise of PD_4 is mainly dominated by $1/f$ noise and g-r noise, which are charge-traps-related [51]. Therefore, although the V_o has been suppressed by reducing the f_{TMIn}, the defects in the In_2O_3 thin films need to be further studied. The spectra density of the noise power can be fitted with a Hooge-type equation:

$$S_n(f) = S_0(\frac{I_d^\beta}{f^\xi}) \tag{7}$$

where S_0 is a constant, β and ξ are two fitting parameters. The calculated ξ of PD_4 is 1.91. The noise current (in) can be estimated by integrating the Hooge-type equation for a given bandwidth, B [52]:

$$i_n^2 = \int_0^B S_n(f)df = \int_0^1 S_n(f)df + \int_1^B S_n(f)df = S_0[\ln(B) + 1] \tag{8}$$

Here, $S_n(f)$ is assumed to be $S_n(1)$ when $f < 1$ Hz. Thus, the i_n of S_4 can be estimated to be 1.68×10^{-6} A $H^{1/2}$. The D* can be expressed as

$$D^* = \frac{(S \cdot B)^{1/2}}{NEP} \tag{9}$$

$$NEP = \frac{i_n}{R} \tag{10}$$

where NEP is the noise equivalent power. Therefore, the D* of PD_4 can be calculated to be 5.43×10^7 cm·$Hz^{1/2}$ W^{-1} (Jones). Figure 7b shows the time-dependent response of PD_4 under period illumination. As shown in Figure 7b, the PD_4 exhibits excellent electric stability after multiple illumination cycles. In addition, the response time of PD_4 is extracted from the time-dependent response curve. The rise time, τ_r, (decay time, τ_d) is defined as the time during which the current rises (decays) from 10% to 90% (90% to 10%). The values of τ_r/τ_d are 0.23 s/0.24 s. Figure 7c shows the time-dependent photo response of S_4 within 30 s, and excellent periodicity can be observed.

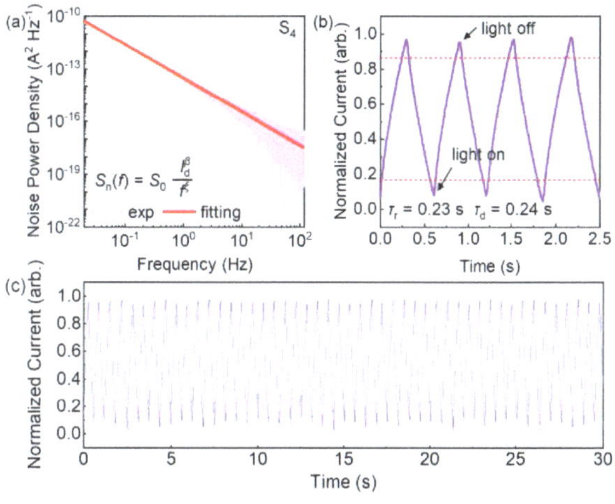

Figure 7. (**a**) Noise power spectra density and (**b**) time-dependent response of PD_4. (**c**) Time-dependent response of S_4 in a 30-s period.

Figure 8a shows the spectra response of the PD_4. As shown in Figure 8a, PD_4 exhibits a peak response in the UVA band. The UV/visible (R_{peak}/R_{400}) rejection ratio can be calculated to be 1241%. The above results indicate that the PD based on In_2O_3 thin film with the f_{TMIn} of 5 sccm mainly works in UVA band. Finally, compared with most of the reported UV PDs, our devices have impressive responsivity as shown in Figure 8b.

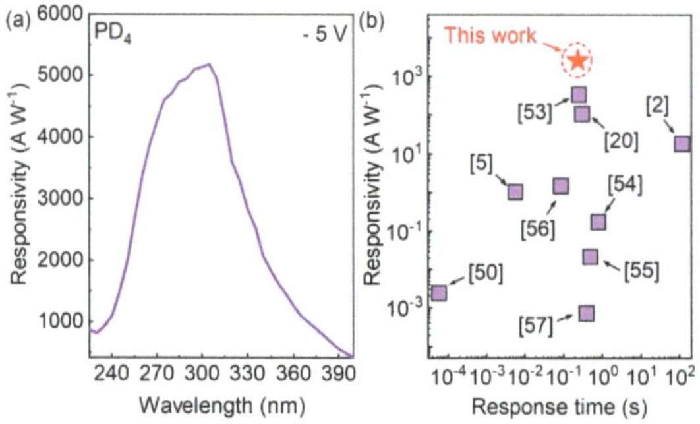

Figure 8. (**a**) Spectra response of PD_4. (**b**) R vs. Response time of this work compared with other ever-reported PDs [2,5,20,50,53–57].

4. Conclusions

In summary, UID-In_2O_3 thin films are grown on sapphire substrates by MOCVD. And the defects such as V_o and background carrier concentration are effectively regulated by changing the f_{TMIn}. With the decrease of the f_{TMIn}, the growth environment gradually becomes oxygen-rich, so that the V_o sates are compensated by O atoms. Consequently, the background carrier concentration decreases from 1.53×10^{19} cm^{-3} to 4.31×10^{16} cm^{-3} with the reduction of the f_{TMIn} from 50 sccm to 5 sccm. And the I_d of the PDs is reduced by more than five orders due to the decrease of the background carrier concentration. In addition, the photo response capacity of the PDs enhances dramatically. The PDCR increases from 0 to 2588 when the f_{TMIn} decreases from 50 sccm to 5 sccm. Finally, a high-performance MSM PD with a high R (2.53×10^3 A W^{-1}), a large EQE ($9383 \times 100\%$), an impressive D* (5.43×10^7 Jones) and a fast response speed (τ_r/τ_d = 0.23 s/0.24 s) is obtained. This work provides an important reference for the realization of high-performance UVA PDs and has great application potential in the field of high-sensitivity UV imaging and communication.

Supplementary Materials: The following supporting information can be downloaded at: https://www.mdpi.com/article/10.3390/cryst14060494/s1, Figure S1: title Cross-section SEM image of In_2O_3 thin film; Figure S2: IV curves of (a) PD_1, (b) PD_2, (c) PD_3 and (d) PD_4 in linear scale.

Author Contributions: Conceptualization, Y.L. and Y.M.; methodology, Y.L. and T.C.; validation, C.Y., Z.Z. and B.Z.; formal analysis, Y.L. and T.C.; investigation, Y.L. and T.C.; resources, X.Z.; data curation, X.Z. and T.C.; writing—original draft preparation, Y.L. and T.C.; writing—review and editing, Y.L. and T.C.; visualization, Y.H., L.Z. and H.Z.; supervision, J.Y. and L.W.; project administration, X.Z. and B.Z.; funding acquisition, X.Z. and B.Z. All authors have read and agreed to the published version of the manuscript.

Funding: This work was funded in part by the National Key Research and Development Program of China under Grants 2021YFB3600202 and 2021YFC2203400; in part by the National Natural Science Foundation of China under Grants 62022081 and 62334010; in part by the Key Laboratory Construction Project of Nanchang under Grant 2020-NCZDSY-008; in part by the Jiangxi Province Double Thousand Plan under Grant S2019CQKJ2638; in part by the Jilin Provincial Science and Technology Department Project under Grant 20220101122JC; in prat by the Changchun University of Science and Technology Project under Grant 6270111102.

Data Availability Statement: The original contributions presented in the study are included in the article/Supplementary Materials, further in-quiries can be directed to the corresponding author.

Acknowledgments: The authors would like to thank Nano Fabrication Facility and Vacuum Interconnected Nanotech Workstation (NANO-X) of Suzhou Institute of Nano-Tech and Nano-Bionics, Chinese Academy of Sciences for their technical support.

Conflicts of Interest: The authors declare no conflict of interest.

References

1. Lemos, S.C.S.; Romeiro, F.C.; de Paula, L.F.; Gonçalves, R.F.; de Moura, A.P.; Ferrer, M.M.; Longo, E.; Patrocinio, A.O.T.; Lima, R.C. Effect of Er^{3+} ions on the phase formation and properties of In_2O_3 nanostructures crystallized upon microwave heating. *J. Solid State Chem.* **2017**, *249*, 58–63. [CrossRef]
2. Li, E.Y.; Zhou, A.F.; Feng, P.X. High-Performance Nanoplasmonic Enhanced Indium Oxide-UV Photodetectors. *Crystals* **2023**, *13*, 689. [CrossRef]
3. Li, Z.Q.; Yan, T.T.; Fang, X.S. Low-dimensional wide-bandgap semiconductors for UV photodetectors. *Nat. Rev. Mater.* **2023**, *8*, 587–603. [CrossRef]
4. Qi, K.; Fu, S.H.; Wang, Y.F.; Han, Y.R.; Fu, R.P.; Gao, C.; Ma, J.A.; Xu, H.Y.; Li, B.S.; Shen, A.D.; et al. High-detectivity solar-blind deep UV photodetectors based on cubic/monoclinic mixed-phase $(In_xGa_{1-x})_2O_3$ thin films. *J. Alloys Compd.* **2023**, *965*, 171473. [CrossRef]
5. Banda, R.R.; Halge, D.I.; Narwade, V.N.; Kaawash, N.M.S.; Thabit, M.Y.H.; Alegaonkar, P.S.; Bogle, K.A. Polarization-independent enhancement in UV photoconductivity of $BiFeO_3$/Sn:In_2O_3 heterostructure. *Phys. B* **2023**, *662*, 414938. [CrossRef]
6. Zhang, S.; Zhang, X.R.; Ren, F.; Yin, Y.; Feng, T.; Song, W.R.; Wang, G.D.; Liang, M.; Xu, J.L.; Wang, J.W.; et al. High responsivity GaN nanowire UVA photodetector synthesized by hydride vapor phase epitaxy. *J. Appl. Phys.* **2020**, *128*, 155705. [CrossRef]
7. Zhang, Y.P.; Zhai, Y.N.; Zhang, H.; Wang, Z.X.; Zhang, Y.F.; Xu, R.L.; Ruan, S.P.; Zhou, J.R. A High-Performance UVA Photodetector Based on Polycrystalline Perovskite $MAPbCl_3/TiO_2$ Nanorods Heterojunctions. *Sensors* **2023**, *23*, 6726. [CrossRef] [PubMed]
8. Mana-ay, H.; Chen, C.S.; Wang, X.H.; Tu, C.S.; Chen, P.Y. Polarization-sensitive UVA photodetector based on heterojunction of ITO and rare-earth doped bismuth ferrite ceramics. *Ceram. Int.* **2022**, *48*, 22083–22095. [CrossRef]
9. Jheng, J.S.; Wang, C.K.; Chiou, Y.Z.; Chang, S.P.; Chang, S.J. MgZnO/SiO_2/ZnO metal–semiconductor–metal dual-band UVA and UVB photodetector with different MgZnO thicknesses by RF magnetron sputter. *Jpn. J. Appl. Phys.* **2020**, *59*, SDFF04. [CrossRef]
10. Wang, B.L.; Ye, L.; Yin, H.; Yu, X.X. Ferroelectrically tuned tunneling photodetector based on graphene/h-BN/In_2Se_3 heterojunction. *Opt. Mater.* **2024**, *150*, 115264. [CrossRef]
11. Chanchal; Jindal, K.; Pandey, A.; Tomar, M.; Jha, P.K. Phase-defined growth of In_2Se_3 thin films using PLD technique for high performance self-powered UV photodetector. *Appl. Surf. Sci.* **2022**, *595*, 153505. [CrossRef]
12. Hase, Y.; Jadhav, Y.; Aher, R.; Sharma, V.; Shah, S.; Punde, A.; Waghmare, A.; Doiphode, V.; Shinde, P.; Rahane, S.; et al. Annealing temperature effect on structural and optoelectronic properties of γ-In_2Se_3 thin films towards highly stable photodetector applications. *J. Mol. Struct.* **2022**, *1265*, 133336. [CrossRef]
13. Li, P.P.; Wang, T.L.; Wang, A.C.; Zhao, L.; Zhu, Y.Q.; Wang, Z.W., Gao, H.L.; Wang, W.J.; Li, K.L.; Du, C.H. Carrier-recirculating broadband photodetector with high gain based on van der Waals In_2Se_3/MoS_2 heterostructure. *Appl. Surf. Sci.* **2024**, *649*, 159135. [CrossRef]
14. Ling, C.C.; Cao, M.; Xue, X.; Zhang, T.; Feng, B.X.; Xue, Q.Z.; Wang, C.K.; Lu, H.P.; Liu, W.P. Large-Scale Synthesis of Vertically Standing In_2S_3 Nanosheets/Pyramidal Silicon Array Heterojunction for Broadband Photodetectors. *Appl. Surf. Sci.* **2023**, *621*, 156901. [CrossRef]
15. Lu, J.T.; Yan, J.H.; Yao, J.D.; Zheng, Z.Q.; Mao, B.J.; Zhao, Y.; Li, J.B. All-Dielectric Nanostructure Fabry-Perot-Enhanced Mie Resonances Coupled with Photogain Modulation toward Ultrasensitive In_2S_3 Photodetector. *Adv. Funct. Mater.* **2021**, *31*, 2007987. [CrossRef]
16. Lu, J.T.; Zheng, Z.Q.; Yao, J.D.; Gao, W.; Zhao, Y.; Xiao, Y.; Li, J.B. 2D In_2S_3 Nanoflake Coupled with Graphene toward High-Sensitivity and Fast-Response Bulk-Silicon Schottky Photodetector. *Small* **2019**, *15*, 1904912. [CrossRef] [PubMed]
17. Zhang, N.N.; Cui, M.Q.; Zhou, J.X.; Shao, Z.T.; Gao, X.Y.; Liu, J.M.; Sun, R.Y.; Zhang, Y.; Li, W.H.; Li, X.H.; et al. High-Performance Self-Powered Photoelectrochemical Ultraviolet Photodetectors Based on an In_2O_3 Nanocube Film. *ACS Appl. Mater. Interfaces* **2024**, *16*, 19167–19174. [CrossRef] [PubMed]
18. Mottram, A.D.; Lin, Y.H.; Pattanasattayavong, P.; Zhao, K.; Amassian, A.; Anthopoulos, T.D. Quasi Two-Dimensional Dye-Sensitized In_2O_3 Phototransistors for Ultrahigh Responsivity and Photosensitivity Photodetector Applications. *ACS Appl. Mater. Interfaces* **2016**, *8*, 4894–4902. [CrossRef] [PubMed]
19. Çaldıran, Z.; Taşyürek, L.B.; Nuhoğlu, Y. The effect of different frequencies and illuminations on the electrical behavior of MoO_3/Si heterojunctions. *J. Mater. Sci. Mater. Electron.* **2021**, *32*, 27950–27961. [CrossRef]
20. Moudgil, A.; Sharma, K.K.; Das, S. In_2O_3/TiO_2 Heterostructure for Highly Responsive Low-Noise Ultraviolet Photodetector. *IEEE Trans. Electron. Dev.* **2020**, *67*, 166–172. [CrossRef]
21. dos Santos, R.B.; Rivelino, R.; Gueorguiev, G.K.; Kakanakova-Georgieva, A. Exploring 2D structures of indium oxide of different stoichiometry. *CrystEngComm* **2021**, *23*, 6661–6667. [CrossRef]

22. Young, S.J.; Liu, Y.H.; Shiblee, M.D.N.I.; Ahmed, K.; Lai, L.T.; Nagahara, L.; Thundat, T.; Yoshida, T.; Arya, S.; Furukawa, H.; et al. Flexible Ultraviolet Photodetectors Based on One-Dimensional Gallium-Doped Zinc Oxide Nanostructures. *ACS Appl. Electron. Mater.* **2020**, *2*, 3522–3529. [CrossRef]
23. Bodur, M.C.; Duman, S.; Orak, I.; Saritas, S.; Baris, O. The photovoltaic and photodiode properties of Au/Carmine/n-Si/Ag diode. *Opt. Laser Technol.* **2023**, *162*, 109251. [CrossRef]
24. Xu, J.; Zheng, W.; Huang, F. Gallium oxide solar-blind ultraviolet photodetectors: A review. *J. Mater. Chem. C* **2019**, *7*, 8753–8770. [CrossRef]
25. Nallabala, N.K.R.; Godavarthi, S.; Kummara, V.K.; Kesarla, M.K.; Saha, D.; Akkera, H.S.; Guntupalli, G.K.; Kumar, S.; Vattikuti, S.V.P. Structural, optical and photoresponse characteristics of metal-insulator-semiconductor (MIS) type Au/Ni/CeO_2/GaN Schottky barrier ultraviolet photodetector. *Mater. Sci. Semicond. Process.* **2020**, *117*, 105190. [CrossRef]
26. Asar, T.; Baran, V.; Kurtulus, G.; Dönmez, M.; Özçelik, H. Platinum doping effect on In_2O_3 MSM IR photodetectors. *Superlattice Microst.* **2018**, *122*, 650–660. [CrossRef]
27. Jesenovec, J.; Weber, M.H.; Pansegrau, C.; McCluskey, M.D.; Lynn, K.G.; McCloy, J.S. Gallium vacancy formation in oxygen annealed β-Ga_2O_3. *J. Appl. Phys.* **2021**, *129*, 245701. [CrossRef]
28. Chen, T.; Zhang, X.; Zhang, L.; Zeng, C.; Li, S.; Yang, A.; Hu, Y.; Li, B.; Jiang, M.; Huang, Z.; et al. High-Speed and Ultrasensitive Solar-Blind Ultraviolet Photodetectors Based on In Situ Grown β-Ga_2O_3 Single-Crystal Films. *ACS Appl. Mater. Interfaces* **2024**, *16*, 6068–6077. [CrossRef]
29. Qian, L.X.; Liu, H.Y.; Zhang, H.F.; Wu, Z.H.; Zhang, W.L. Simultaneously improved sensitivity and response speed of β-Ga_2O_3 solar-blind photodetector via localized tuning of oxygen deficiency. *Appl. Phys. Lett.* **2019**, *114*, 5088665. [CrossRef]
30. Ge, C.Y.; Liu, Z.Y.; Zhu, Y.C.; Zhou, Y.L.; Jiang, B.R.; Zhu, J.X.; Yang, X.K.; Zhu, Y.X.; Yan, S.Y.; Hu, H.J.; et al. Insight into the High Mobility and Stability of In_2O_3:H Film. *Small* **2024**, *20*, 2304721. [CrossRef]
31. Qian, H.; Zhang, X.D.; Ma, Y.J.; Zhang, L.; Chen, T.W.; Wei, X.; Tang, W.B.; Zhou, X.; Feng, B.Y.; Fan, Y.M.; et al. Quasi-vertical ε-Ga_2O_3 solar-blind photodetectors grown on p-Si substrates with Al_2O_3 buffer layer by metalorganic chemical vapor deposition. *Vacuum* **2022**, *200*, 111019. [CrossRef]
32. Das, S.; Paikaray, S.; Swain, I.; Senapati, S.; Naik, R. Tuning in linear and nonlinear optical parameters by interfacial mixing of Sb/Ag_2Se bilayer thin films under annealing at different temperatures for optoelectronic applications. *Surf. Interfaces* **2023**, *42*, 103395. [CrossRef]
33. Li, M.Q.; Yang, N.; Wang, G.G.; Zhang, H.Y.; Han, J.C. Highly preferred orientation of Ga_2O_3 films sputtered on SiC substrates for deep UV photodetector application. *Appl. Surf. Sci.* **2019**, *471*, 694–702. [CrossRef]
34. Wu, N.; Wang, C.; Slattum, P.M.; Zhang, Y.Q.; Yang, X.M.; Zang, L. Persistent Photoconductivity in Perylene Diimide Nanofiber Materials. *ACS Energy Lett.* **2016**, *1*, 906–912. [CrossRef]
35. Qian, L.X.; Wu, Z.H.; Zhang, Y.Y.; Lai, P.T.; Liu, X.Z.; Li, Y.R. Ultrahigh-Responsivity, Rapid-Recovery, Solar-Blind Photodetector Based on Highly Nonstoichiometric Amorphous Gallium Oxide. *ACS Photonics* **2017**, *4*, 2203–2211. [CrossRef]
36. Greczynski, G.; Hultman, L. Reliable determination of chemical state in x-ray photoelectron spectroscopy based on sample-work-function referencing to adventitious carbon: Resolving the myth of apparent constant binding energy of the C 1s peak. *Appl. Surf. Sci.* **2018**, *451*, 99–103. [CrossRef]
37. Hu, Y.; Zhang, L.; Chen, T.W.; Ma, Y.J.; Tang, W.B.; Huang, Z.J.; Li, B.T.; Xu, K.; Mudiyanselage, D.H.; Fu, H.Q.; et al. High-performance ε-Ga_2O_3 solar-blind ultraviolet photodetectors on Si (100) substrate with molybdenum buffer layer. *Vacuum* **2023**, *213*, 112130. [CrossRef]
38. Ma, Y.J.; Feng, B.Y.; Zhang, X.D.; Chen, T.W.; Tang, W.B.; Zhang, L.; He, T.; Zhou, X.; Wei, X.; Fu, H.Q.; et al. High-performance β-Ga_2O_3 solar-blind ultraviolet photodetectors epitaxially grown on (110) TiO_2 substrates by metalorganic chemical vapor deposition. *Vacuum* **2021**, *191*, 110402. [CrossRef]
39. Wang, Y.C.; Zhu, L.; Liu, Y.; Vovk, E.I.; Lang, J.Y.; Zhou, Z.X.; Gao, P.; Li, S.G.; Yang, Y. Understanding surface structures of In_2O_3 catalysts during CO_2 hydrogenation reaction using time-resolved IR, XPS with in situ treatment, and DFT calculations. *Appl. Surf. Sci.* **2023**, *631*, 157534. [CrossRef]
40. Ma, Q.; Zheng, H.M.; Shao, Y.; Zhu, B.; Liu, W.J.; Ding, S.J.; Zhang, D.W. Atomic-Layer-Deposition of Indium Oxide Nano-films for Thin-Film Transistors. *Nanoscale Res. Lett.* **2018**, *13*, 4. [CrossRef]
41. Peng, Y.H.; He, C.C.; Zhao, Y.J.; Yang, X.B. Multi-peak emission of In_2O_3 induced by oxygen vacancy aggregation. *J. Appl. Phys.* **2023**, *133*, 075702. [CrossRef]
42. Bierwagen, O.; Speck, J.S. High electron mobility In_2O_3 (001) and (111) thin films with nondegenerate electron concentration *Appl. Phys. Lett.* **2010**, *97*, 072103. [CrossRef]
43. Chen, A.; Zhu, K.G.; Zhong, H.C.; Shao, Q.Y.; Ge, G.L. A new investigation of oxygen flow influence on ITO thin films by magnetron sputtering. *Sol. Energy Mater. Sol. Cells* **2014**, *120*, 157–162. [CrossRef]
44. Heinemann, M.D.; Berry, J.; Teeter, G.; Unold, T.; Ginley, D. Oxygen deficiency and Sn doping of amorphous Ga_2O_3. *Appl. Phys. Lett.* **2016**, *108*, 022107. [CrossRef]
45. Carrano, J.C.; Li, T.; Grudowski, P.A.; Eiting, C.J.; Dupuis, R.D.; Campbell, J.C. Comprehensive characterization of metal-semiconductor-metal ultraviolet photodetectors fabricated on single-crystal GaN. *J. Appl. Phys.* **1998**, *83*, 6148–6160. [CrossRef]
46. Hellings, G.; John, J.; Lorenz, A.; Malinowski, P.; Mertens, R. AlGaN Schottky Diodes for Detector Applications in the UV Wavelength Range. *IEEE Trans. Electron. Dev.* **2009**, *56*, 2833–2839. [CrossRef]

47. Huang, H.L.; Xie, Y.N.; Zhang, Z.F.; Zhang, F.; Xu, Q.; Wu, Z.Y. Growth and fabrication of sputtered TiO_2 based ultraviolet detectors. *Appl. Surf. Sci.* **2014**, *293*, 248–254. [CrossRef]
48. Abdullah, Q.N.; Yam, F.K.; Mohmood, K.H.; Hassan, Z.; Qaeed, M.A.; Bououdina, M.; Almessiere, M.A.; Al-Otaibi, A.L.; Abdulateef, S.A. Free growth of one-dimensional β-Ga_2O_3 nanostructures including nanowires, nanobelts and nanosheets using a thermal evaporation method. *Ceram. Int.* **2016**, *42*, 13343–13349. [CrossRef]
49. Nikitskiy, I.; Goossens, S.; Kufer, D.; Lasanta, T.; Navickaite, G.; Koppens, F.H.L.; Konstantatos, G. Integrating an electrically active colloidal quantum dot photodiode with a graphene phototransistor. *Nat. Commun.* **2016**, *7*, 11954. [CrossRef]
50. Abbas, S.; Ban, D.K.; Kim, J. Functional interlayer of In_2O_3 for transparent SnO_2/SnS_2 heterojunction photodetector. *Sens. Actuators A-Phys.* **2019**, *293*, 215–221. [CrossRef]
51. Fang, Y.J.; Huang, J.S. Resolving Weak Light of Sub-picowatt per Square Centimeter by Hybrid Perovskite Photodetectors Enabled by Noise Reduction. *Adv. Mater.* **2015**, *27*, 2804. [CrossRef] [PubMed]
52. Chen, Q.; Yang, J.W.; Osinsky, A.; Gangopadhyay, S.; Lim, B.; Anwar, M.Z.; Khan, M.A.; Kuksenkov, D.; Temkin, H. Schottky barrier detectors on GaN for visible-blind ultraviolet detection. *Appl. Phys. Lett.* **1997**, *70*, 2277–2279. [CrossRef]
53. Pooja, P.; Chinnamuthu, P. Annealed n-TiO_2/In_2O_3 nanowire metal-insulator-semiconductor for highly photosensitive low-noise ultraviolet photodetector. *J. Alloys Compd.* **2021**, *854*, 157229. [CrossRef]
54. Zhang, M.X.; Yu, H.; Li, H.; Jiang, Y.; Qu, L.H.; Wang, Y.X.; Gao, F.; Feng, W. Ultrathin In_2O_3 Nanosheets toward High Responsivity and Rejection Ratio Visible-Blind UV Photodetection. *Small* **2023**, *19*, 2205623. [CrossRef] [PubMed]
55. Cui, M.Q.; Shao, Z.T.; Qu, L.H.; Liu, X.; Yu, H.; Wang, Y.X.; Zhang, Y.X.; Fu, Z.D.; Huang, Y.W.; Feng, W. MOF-Derived In_2O_3 Microrods for High-Performance Photoelectrochemical Ultraviolet Photodetectors. *ACS Appl. Mater. Interfaces* **2022**, *14*, 39046–39052. [CrossRef] [PubMed]
56. Sharmila, B.; Dwivedi, P. In_2O_3 decorated TiO_2 for broadband photosensing applications. *Semicond. Sci. Technol.* **2023**, *38*, 115009.
57. Veeralingam, S.; Badhulika, S. Enhanced carrier separation assisted high-performance piezo-phototronic self-powered photodetector based on core-shell $ZnSnO_3@In_2O_3$ heterojunction. *Nano Energy* **2022**, *98*, 107354. [CrossRef]

crystals

Article

Influence of the Incorporation of Nd in ZnO Films Grown by the HFCVD Technique to Enhance Photoluminiscence Due to Defects

Marcos Palacios Bonilla [1,*], Godofredo García Salgado [1,*], Antonio Coyopol Solís [1], Román Romano Trujillo [1], Fabiola Gabriela Nieto Caballero [2], Enrique Rosendo Andrés [1], Crisóforo Morales Ruiz [1], Justo Miguel Gracia Jiménez [3] and Reina Galeazzi Isasmendi [1]

1 Centro de Investigación en Dispositivos Semiconductores, Benemérita Universidad Autónoma de Puebla, Ed. IC5, Av. San Claudio y Blvd. 18 Sur, Col. San Manuel, Puebla 72570, Mexico; antonio.coyopol@correo.buap.mx (A.C.S.); roman.romano@correo.buap.mx (R.R.T.); enrique.rosendo@correo.buap.mx (E.R.A.); crisoforo.morales@correo.buap.mx (C.M.R.); reina.galeazzi@correo.buap.mx (R.G.I.)

2 Facultad de Ciencias Químicas, Benemérita Universidad Autónoma de Puebla, Ed. FCQ4, Av. San Claudio y Blvd. 18 Sur, Col. San Manuel, Puebla 72570, Mexico; fabiola.nieto@correo.buap.mx

3 Instituto de Física, Benemérita Universidad Autónoma de Puebla, Av. San Claudio y Blvd. 18 Sur, Col. San Manuel, Puebla 72570, Mexico; gracia@ifuap.buap.mx

* Correspondence: marcos.palaciosbonilla@alumno.buap.mx (M.P.B.); godofredo.garcia@correo.buap.mx (G.G.S.)

Abstract: In this work, optical–structural and morphological behavior when Nd is incorporated into ZnO is studied. ZnO and Nd-doped ZnO (ZnO-Nd) films were deposited at 900 °C on Silicon n-type substrates (100) by using the Hot Filament Chemical Vapor Deposition (HFCVD) technique. For this, pellets were made by from powders of ZnO(s) and a mixture of ZnO(s):Nd(OH)$_3$(s). The weight percent of the mixture ZnO:Nd(OH)$_3$ in the pellet is 1:3. The gaseous precursor generation was carried out by chemical decomposition of the pellets using atomic hydrogen which was produced by a tungsten filament at 2000 °C. For the ZnO film, diffraction planes (100), (002), (101), (102), (110), and (103) were found by XRD. For the ZnO-Nd film, its planes are displaced, indicating the incorporation of Nd into the ZnO. EDS was used to confirm the Nd in the ZnO-Nd film with an atomic concentration (at%) of Nd = 10.79. An improvement in photoluminescence is observed for the ZnO-Nd film; this improvement is attributed to an increase in oxygen vacancies due to the presence of Nd. The important thing about this study is that by the HFCVD method, ZnO-Nd films can be obtained easily and with very short times; in addition, some oxide compounds can be obtained individually as initial precursors, which reduces the cost compared to other techniques. Something interesting is that the incorporation of Nd into ZnO by this method has not yet been studied, and depending on the method used, the PL of ZnO with Nd can increase or decrease, and by the HFCVD method the PL of the ZnO film, when Nd is incorporated, increases more than 15 times compared to the ZnO film.

Keywords: HFCVD; silicon; ZnO-Nd; sea urchin

Citation: Palacios Bonilla, M.; García Salgado, G.; Coyopol Solís, A.; Romano Trujillo, R.; Nieto Caballero, F.G.; Rosendo Andrés, E.; Morales Ruiz, C.; Gracia Jiménez, J.M.; Galeazzi Isasmendi, R. Influence of the Incorporation of Nd in ZnO Films Grown by the HFCVD Technique to Enhance Photoluminiscence Due to Defects. *Crystals* **2024**, *14*, 491. https://doi.org/10.3390/cryst14060491

Academic Editors: Miłosz Grodzicki, Damian Wojcieszak and Michał Mazur

Received: 25 April 2024
Revised: 14 May 2024
Accepted: 16 May 2024
Published: 23 May 2024

1. Introduction

ZnO presents n-type electrical conductivity and has a wide direct bandgap of ~3.37 eV at room temperature [1,2]. In addition, it presents interesting properties such as: high chemical and thermal stability; good optical and electrical properties; high transparency in the Vis/near-IR spectral region [3]; and a large exciton binding energy of 60 mV [2,4–6]. Such properties have allowed ZnO to be used in optoelectronic device applications such as: gas sensors [7], solar cells [8], ultraviolet (UV) detectors [9], light-emitting diodes, and lasers [10]. On the other hand, ZnO presents a better optical response in PL intensity and a low band gap when it is doped and forms compounds with different chemical elements, for example Nd [11], Ni [12], and Ta [13]. Doping ZnO with metal atoms such as Nd introduces new properties in ZnO that are useful in different areas such as photocatalysis and optoelectronics [14,15]. Nd-doped ZnO films enhance PL intensity at deep level emissions (~511 nm) [11]. Some methods such as spray pyrolysis [14], pulsed electron beam [16], and sol-gel [17], have been used to obtain Nd-doped ZnO films. However, using the HFCVD technique for this process has not been explored. The HFCVD technique allows the obtention of high deposit rates, so long times for film growth are not necessary. In addition, solid sources have been used for the formation of gaseous precursors for the deposit of materials such as: ZnO [18], ZnO-Ni [12], and ZnO-Ta [13]. Therefore, the study of ZnO with the incorporation of Nd from a solid pellet based on a mixture of ZnO(s):$Nd(OH)_3$(s), attacked with atomic hydrogen (H°), is feasible. The atomic hydrogen is produced by using a tungsten filament at 2000 °C, etching the solid source (pellet) to obtain chemical precursors that diffuse and deposit on a Si substrate [13,19]. The HFCVD process has been reported in other works [18,20]. Using the HFCVD technique, sphere-shaped structures of ZnO can be grown, which have a large surface–volume ratio which is important for devices such as drug delivery devices [21] and gas sensors [18], and in solar energy conversion and field emission [18,22]. Thus, the objective of this work is to study the structural, morphological, and optical properties of ZnO and compare them with ZnO-Nd films on silicon substrates.

2. Experimental Details

2.1. Pellets Preparation

A ZnO(s):$Nd(OH)_3$(s) pellet (0.4 g) was prepared by compressing ZnO(s) powder (Sigma-Aldrich CAS-No:1314-13-2, Estado de México, México) combined with $Nd(OH)_3$(s) powder (Sigma-Aldrich CAS-No:16469-17-3). The weight percent of the mixture ZnO(s):$Nd(OH)_3$(s) in the pellet was 1:3. The reason that the pellet was in that proportion is because the amount of Nd that can be released from the ZnO:$Nd(OH)_3$ pellet in the film growth process is minimal (less than 1% in weight percentage). In addition, the 1:3 ratio was prepared because other samples were made and it was observed that the PL intensity increases as the Nd in the ZnO increases, and this sample is very interesting and high in PL intensity. The quantities used in each pellet of this work are shown in Table 1.

Table 1. Experimental conditions used for ZnO and ZnO-Nd films.

Film	Pellet	Proportion in Weight Percentage ZnO:$Nd(OH)_3$	ZnO Weight (g)	$Nd(OH)_3$ Weight (g)	Substrate Temperature (°C)	Process Time (min.)
ZnO	ZnO	1:0	0.4	0	900	3
ZnO-Nd	ZnO:$Nd(OH)_3$	1:3	0.1	0.3	900	3

2.2. Growth Process of ZnO and ZnO-Nd Films

Silicon n-types with orientation (100) and resistivity 1–20 Ω·cm were used as substrates. Firstly, Si-substrates were cleaned following a process in an ultrasonic bath for 10 min in each chemical solution, including: xylene, acetone, and methanol. Additionality, Si substrates were immersed for 1 min in HF (10%) to remove native oxide. For the film growth process, H_2 flow (50 sccm) was incorporated in reaction chamber (Figure 1a); afterwards, atomic hydrogen (H°) was produced by a tungsten filament at 2000 °C., although the growth temperature used for deposits was 900 °C for 3 min. The generation of gaseous precursors occured through the reaction of a solid source (pellet based on a mixture of ZnO(s) and $Nd(OH)_3$(s)) with atomic hydrogen (Figure 1b).

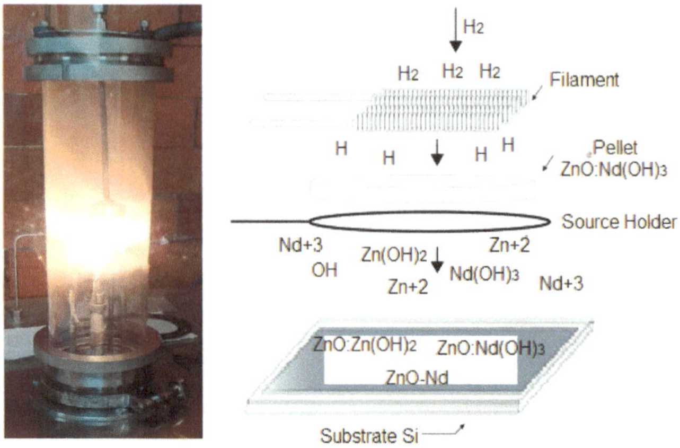

(**a**) (**b**)

Figure 1. (**a**) HFCVD reactor and (**b**) internal configuration of HFCVD reactor.

Regarding the reaction for ZnO films grown using the HFCVD technique, it has already been reported in other works [23]. On the other hand, for the ZnO-Nd film grown in this work, the substrate chemical reactions are presented as follows:

$$ZnO(s) + Nd(OH)_3(s) + H(g) \rightarrow Zn(g) + Nd(g) + 4OH(g) \quad (1)$$

$$4Zn(g) + Nd(g) + 4OH(g) \rightarrow Zn(s) + ZnO(s) + Zn(OH)_2(s) + ZnONd(s) + H_2(g) \quad (2)$$

Finally, XRD, SEM, EDS, and PL measurements were performed. The structural characterization was carried out with an X-ray diffractor (BRUKER D8 with Cu Kα radiation (λ = 1.541 Å), Billerica, MA, USA). The surface morphology and elements of the samples were characterized by using a Jeol JSM-7800F (JEOL Ltd., Tokyo, Japan) field emission scanning electron microscope equipped with an Oxford Instrument X-Max spectrometer for elemental analysis (JEOL Ltd., Tokyo, Japan). Photoluminescence measurements were performed at room temperature with a Horiba Jobin Yvon NanoLog FR3 UV-Vis-NIR fluorescence spectrometer (Horiba Ltd., Kyoto, Japan).

3. Results and Discussion

3.1. Structural Characterization by X-ray Diffraction (XRD) Analysis

ZnO and ZnO-Nd films were analyzed using XRD. Figure 2 shows the pattern of XRD for the samples measured from 25° to 65° in the 2θ mode. For the ZnO film in the (a) pattern, three Zn planes (100), (101), and (102) located at 39.16°, 43.38° and 54.42° are observed, which match with the Zn hexagonal structure of the pdf card 004-0831. It is observed that the planes (100), (002), (101), (102), (110), and (103) of ZnO which are located at 31.98°, 34.6°, 36.44°, 47.74°, 56.78°, and 63.08° match according to the wurtzite ZnO structure of the pdf card 075-0576. In pattern (b), for the ZnO-Nd film the same ZnO film planes are observed. Furthermore, it can be observed that the ZnO-Nd film planes shift slightly to the left compared to the ZnO film planes, and it is suggested that this slight shift is due to the change in lattice parameters; however, the wurtzite structure of the crystal remains unchanged [24]. So, for this reason it is suggested that Nd is incorporated into the ZnO matrix. This shift is due to the difference in the ionic radii of Zn^{+2} and Nd^{+3}. Since the ionic radius of Nd^{+3} (0.983 Å) is larger than that of Zn^{+2} (0.74 Å), an expansive stress is created in the lattice; therefore, it is suggested that there is an expansion in the ZnO lattice resulting in the displacement of the ZnO-Nd planes and it is suggested that this expansion in the crystal lattice confirms that Nd is introduced substitutionally into the ZnO lattice [15,24].

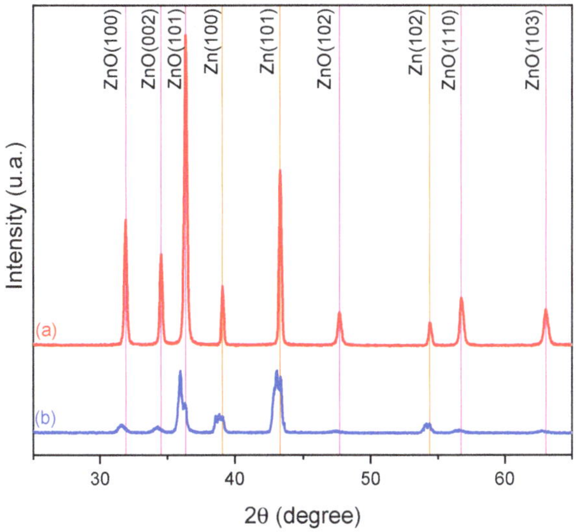

Figure 2. X-ray diffraction patterns for (a) ZnO and (b) ZnO-Nd films.

Figure 3 analyzes Figure 2 better: the planes of ZnO, Zn, $Zn(OH)_2$, and $Nd(OH)_3$ are shown with different symbols. In the ZnO film (Figure 3a), the ZnO and Zn planes are observed, so the core–shell structures are observed, where the metallic Zn is the core and the ZnO is the shell [12,13]. In addition, the $Zn(OH)_2$ planes [25] are observed because in the film growth process some Zn^{+2} joins with the OH that is being transported, forming $Zn(OH)_2$; then, some Zn(OH) is also deposited on the surface of the film, remaining in the form of ZnO:$Zn(OH)_2$ composites. In the ZnO-Nd film (Figure 3b), the same planes of the ZnO film are observed; however, some $Nd(OH)_3$ planes [26] also appear. This is because in the film growth process some Nd^{+3} atoms join with the OH that is being transported, forming $Nd(OH)_3$, and for this reason some $Nd(OH)_3$ is also deposited on the surface of the film, remaining in the form of ZnO:$Nd(OH)_3$ composites. In addition, the $Zn(OH)_2$ planes [25] are observed, so there are also some ZnO:$Zn(OH)_2$ composites on the surface of the film.

To calculate the lattice parameters, Equation (3) is used:

$$\frac{1}{d^2_{(hkl)}} = \frac{4}{3a^2}\left(h^2 + k^2 + hk\right) + \frac{l^2}{c^2} \tag{3}$$

In this way, the interplanar spacing $d_{(hkl)}$ can be obtained from the relation $\frac{1}{d_{hkl}} = \frac{2\sin\theta}{\lambda}$, where h,k,l are the miller indices of the respective plane, and a and c are the lattice parameters which are calculated with the (100), and (002) planes at 31.98° and 34.6° [12]. Table 2 shows the lattice parameters for the (002) plane of the different films obtained in this work.

Table 2. Lattice parameters of ZnO and ZnO-Nd films of plane (002).

Sample	a (Å)	c (Å)	c/a
ZnO	3.2289	5.1806	1.6044
ZnO-Nd	3.2547	5.2187	1.6034

It can be seen in Table 2 that the parameter "c" increases and the parameter "a" increases, and so for these reasons, it is suggested that Nd^{+3} is incorporated into the ZnO lattice in a substitutional manner, and it is proposed that Nd^{+3} ions replace Zn^{+2} ions, thus obtaining Nd-doped ZnO [24].

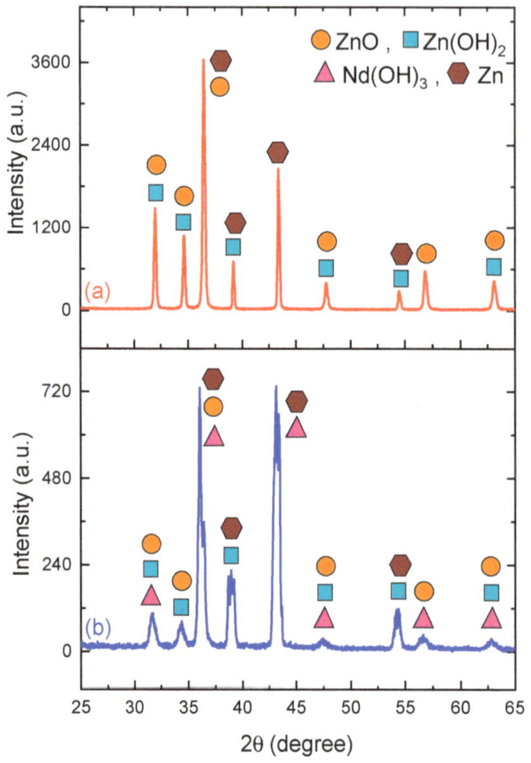

Figure 3. A better analysis of Figure 2 is shown for the XRD patterns of (**a**) ZnO and (**b**) ZnO-Nd films.

3.2. Scanning Electron Microscope (SEM)

SEM measurements were made to observe the morphology of the ZnO and ZnO-Nd films. Figure 4a,b shows the ZnO film and Figure 4c–f shows the ZnO-Nd film at different zoom levels.

SEM images in Figure 4a,b, for a ZnO film, show sphere-type structures on the surface of the film. It is suggested that the growth process of sphere-type structures occurs when molecular hydrogen enters into the reactor through the filament at 2000 °C and decomposes in atomic hydrogen, which attacks the ZnO(s) pellet, forming the precursors Zn(g) and OH(g) as proposed in [13,27], starting with the following chemical reactions:

$$ZnO(s) + H(g) \rightarrow Zn(g) + OH(g) \quad (4)$$

$$3Zn(g) + 4OH(g) \rightarrow Zn(s) + ZnO(s) + Zn(OH)_2(s) + H_2O(g) \quad (5)$$

Molecular hydrogen decomposes into atomic hydrogen due to the temperature of the filament. The atomic H releases Zn and O atoms from the ZnO pellet, according to reaction 4. In reaction 5, core–shell (Zn/ZnO) structures are formed on the silicon substrate; this is deduced from the XRD planes of Zn and ZnO [12,13]. Figure 4b shows the surface of an individual ZnO sphere (Figure 4a) with a zoom of ×1000. Figure 4c–f show the ZnO-Nd film. The formation process of ZnO-Nd spheres is similar to that of ZnO spheres; the difference is that ZnONd structures are formed in the former. It can be observed the sphere-like structure that it has on its surface is like wires; this is suggested to be because the Nd atoms act as a catalyst and nucleate, resulting in a sea urchin-like structure [28]. Furthermore, it can be observed that when Nd is incorporated, there is a decrease in the size of the spheres. It is observed that the spheres on the ZnO-Nd film have a smaller size than in the ZnO film and it is suggested that the decrease in the size of the ZnO-Nd spheres is due to the incorporation of Nd. That is to say, it is due to the nucleation mechanism in the growth process; because the ionic radius of Nd^{+3} is greater than that of Zn^{+2}, it has a lower nucleation rate than in the ZnO film which results in a smaller sphere size [29]. From Figure 4b,d ($\times$1000), it is estimated that the ZnO spheres are larger, since in the ZnO sphere (Figure 4b) only part of the surface of an individual ZnO sphere can be observed, and Figure 4d shows complete ZnO-Nd spheres because they are smaller in size. Figure 5 shows the distribution of the different chemical elements for the ZnO-Nd film. Figure 6 shows each chemical element individually. The film appears relatively homogeneous; however, it can be seen that there is an agglomeration of many Nd atoms (as also seen in Figure 4f), which suggests that there is a ZnO:Nd$(OH)_3$ composite because in XRD the Nd$(OH)_3$ planes for the ZnO-Nd film are observed.

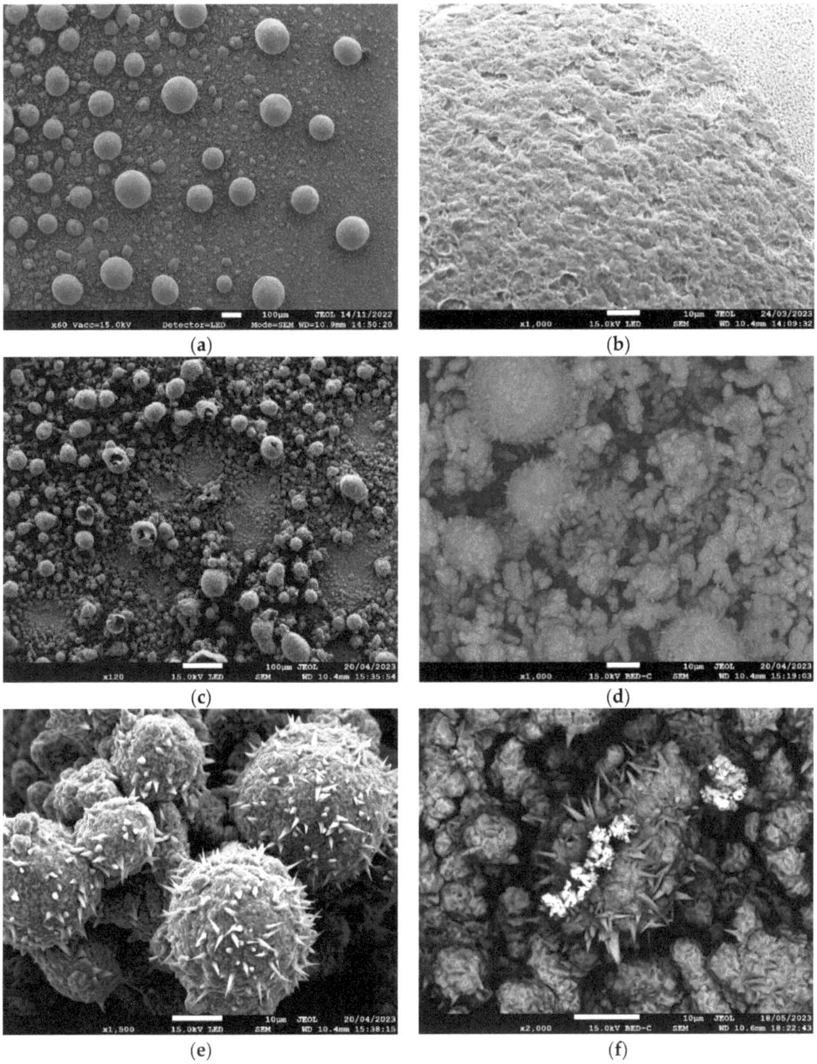

Figure 4. SEM for (**a**,**b**) ZnO and (**c**–**f**) ZnO-Nd films at different zoom levels.

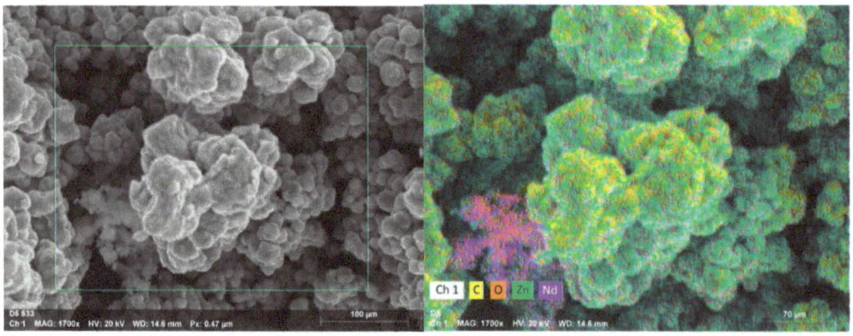

Figure 5. Color mapping of the different chemical elements of the ZnO-Nd film.

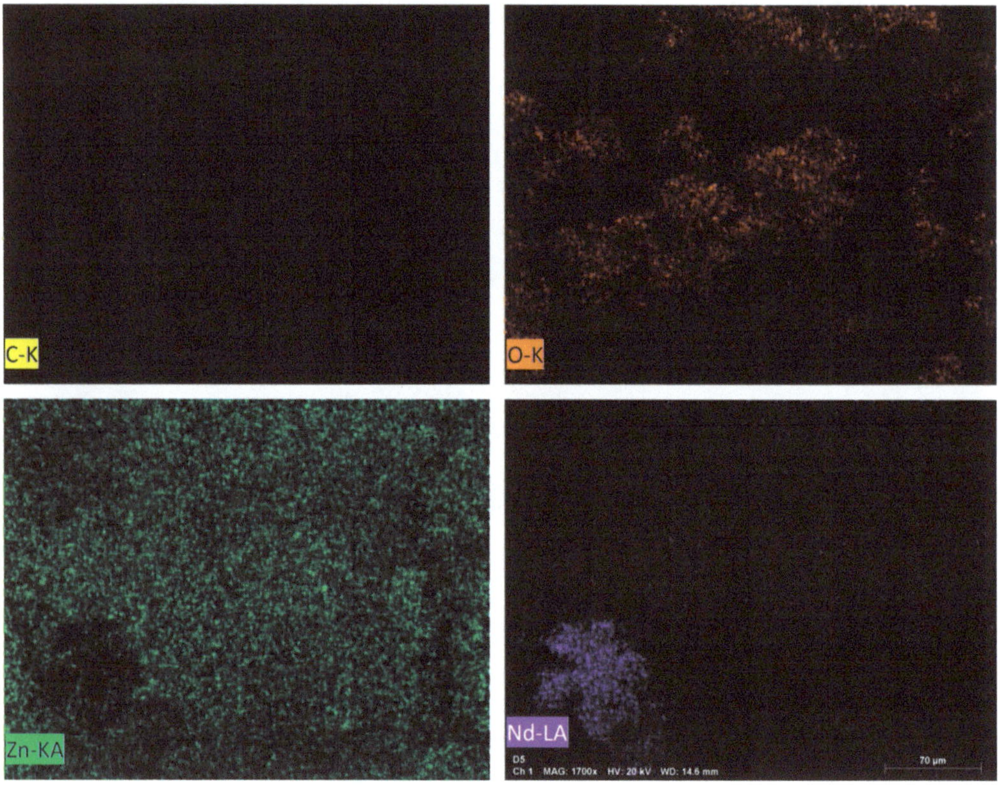

Figure 6. Chemical elements individually for the ZnO-Nd film.

3.3. *Energy Dispersive X-ray (EDS) Analysis*

Figure 7a shows the EDS analysis for the ZnO film. In the EDS of the ZnO film, it is observed that the film presents the chemical elements of Zn and O. For this reason, it is suggested that when the H impacts the ZnO pellet, it decomposes it into Zn and O atoms that begin to be transported into the reactor. Subsequently, the Zn, due to its boiling point (907 °C), adheres to the substrate (900 °C) due to the temperature difference and remains as a nucleation site, but since the O atoms are also being transported, they begin to bind to the Zn and, thus, the ZnO film is obtained [13]. The carbon peak corresponds to the piece of paper used in the SEM measurement. Figure 7b shows the EDS of the ZnO-Nd film, where the presence of Nd is confirmed. It is observed that the film presents the chemical elements of Zn, O, and Nd. In the ZnO-Nd film, something very similar happens: the H impacts the ZnO:Nd$(OH)_3$ pellet and decomposes it into Zn, O, and Nd atoms that begin to be transported inside the reactor; in this way, ZnO is formed as described above, but since Nd atoms are also being transported here, they begin to bind to the O of ZnO and, thus, the ZnO-Nd film is obtained [13]. Table 3 shows the different compositions of the ZnO and ZnO-Nd films.

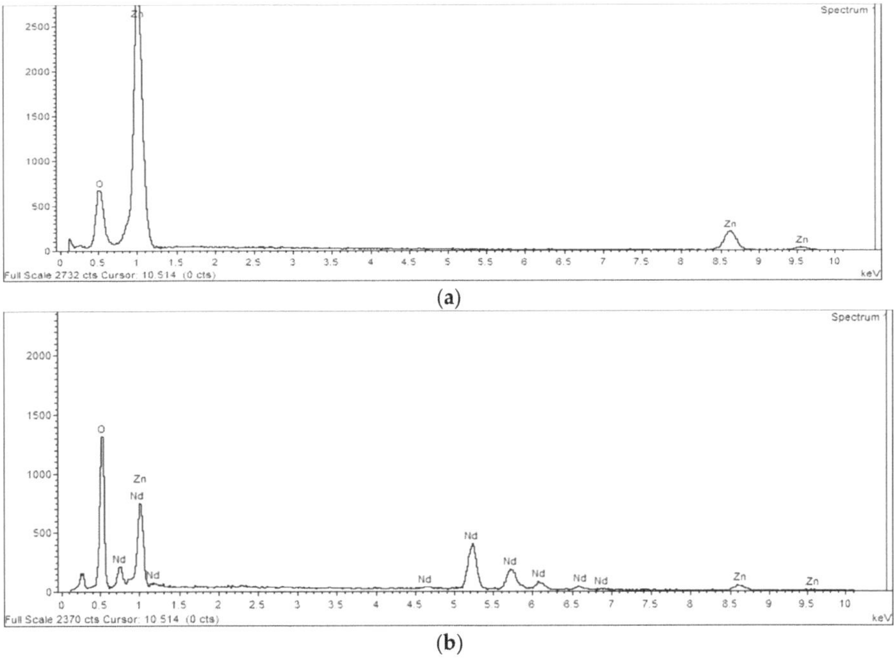

Figure 7. (**a**) EDS for ZnO film. (**b**) EDS for ZnO-Nd film.

Table 3. Average atomic percentage of elemental composition of ZnO and ZnO-Nd films.

Sample	Zn (%)	O (%)	Nd (%)
ZnO	69.36	30.64	0
ZnO-Nd	43.99	45.22	10.79

It can be seen that the ZnO film has an atomic percentage of approximately Zn = 69.36%, which suggests that it is a film rich in zinc. It can be observed in the ZnO-Nd film (Figure 4f) that there is approximately an atomic percentage of Nd = 10.79%; therefore, the presence of Nd is confirmed. For Figure 4e, there is an average atomic percentage of Nd = 1.25%, Zn = 63.64%, and O = 35.11% so it is suggested that there is doping; however, in Figure 4f it can be seen that an agglomerate of Nd remained on the surface of the film with Nd = 10.79% (Table 3), so it is suggested that it is also a ZnO:Nd$(OH)_3$ composite, and this can be seen better in Figure 6.

3.4. Photoluminescence Analysis (PL)

Figure 8 shows the PL spectra of the ZnO and ZnO-Nd films at room temperature measured from 350 to 750 nm, with the purpose of studying the behavior of Nd in the ZnO.

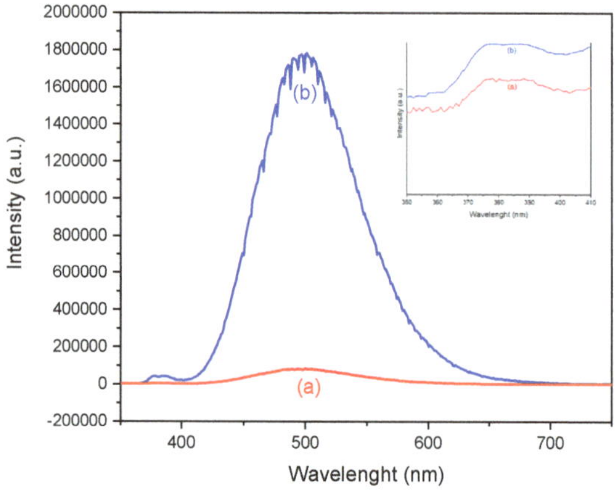

Figure 8. Photoluminescence spectra of (a) ZnO and (b) ZnO-Nd films. A PL graph is also presented with a zoom in the NBE region of ~380 nm.

For the ZnO film, a band ~380 nm named as near-band-edge emission NBE is observed, and this is due to the band-to-band transition [13]; the band ~502 nm, called the deep level emission which is abbreviated to DLE, is due to the transitions through deep centers with energy levels in the forbidden gap of the hydroxyl group, and is also due to different defects in the lattice, such as oxygen or zinc interstitials, oxygen or zinc vacancies, or external impurities due to the substitution of atoms [13,29,30]. For the ZnO-Nd film, an increase in the intensity of the DLE band is observed. It is suggested that this is because when Nd is incorporated into the ZnO, more defect centers such as oxygen vacancies are created and for this reason a PL increase occurs [31]. In addition, due to the distortion of the lattice as mentioned in XRD, and other defects, it is suggested that the energy gaps are split and for this reason a greater wavelength range is covered in the PL spectrum [32]. From SEM, it can be observed that there is a decrease in the size of the spheres, and because it is reported in the literature that, when there is a decrease in particle size, more oxygen vacancies are created, it is suggested that for this reason the PL also increases [30]. Furthermore, in Figure 8 for the ZnO and ZnO-Nd films, in the NBE band region, two different bands are observed that correspond to ZnO (~380 nm) and $Zn(OH)_2$ (~387 nm) [33]. The two bands, NBE and DLE, of $Zn(OH)_2$ are similar to the two bands (NBE and DLE) of ZnO, only the intensity of the PL varies a little [33]. Based on the PL results and because the $Zn(OH)_2$ planes are observed in XRD, it is suggested that there are also some ZnO:$Zn(OH)_2$ composites on the surface of the ZnO and ZnO-Nd films.

4. Conclusions

The objective of this research is to observe how $Nd(OH)_3$ behaves when reacting with ZnO using the HFCVD method, since ZnO-Nd formed by this technique has not yet been explored; in addition, with growth by HFCVD, ZnO and ZnO-Nd films are obtained relatively easily and with short times. Therefore, the structural, morphological, and optical properties of ZnO and ZnO-Nd films on silicon substrates are studied. From XRD studies, it is observed that the ZnO planes match with the hexagonal wurtzite structure of ZnO. It is also observed in the diffractogram that there is a displacement of the planes of the ZnO-Nd film, which is due to the fact that the Nd is incorporated into the ZnO, thus achieving doping. Furthermore, the planes of Zn, ZnO, $Zn(OH)_3$, and $Nd(OH)_3$ are observed individually, so the presence of a ZnO:$Nd(OH)_3$ composite is also noted. From the SEM results, the different morphology observed in the sphere-like structures of the ZnO-Nd film is due to the fact that the Nd atoms act as a catalyst and nucleate, resulting in a structure similar to that of a sea urchin. It is also observed that the sphere-like structures in the ZnO-Nd film have a smaller size than in the ZnO film; this is due to the incorporation of Nd in the ZnO. In addition, from the different morphology and decrease in size of the ZnO-Nd spheres, it is deduced that the Nd is doping the ZnO; however, as some agglomerations of $Nd(OH)_3$ are also observed on the surface of the film, it is deduced that it is also a ZnO:$Nd(OH)_3$ composite. EDS analysis confirms the presence of Nd. Furthermore, in the ZnO-Nd sphere there is a low average atomic percentage of Nd = 1.25%, so it can be deduced that there is doping; however, in the $Nd(OH)_3$ agglomeration, there is a higher average atomic percentage of Nd = 10.79%, andso it follows that there is also a ZnO:$Nd(OH)_3$ composite. From the PL measurement, when Nd is incorporated into ZnO as a dopant, it causes an increase in O vacancies, thus achieving an increase in the intensity of photoluminescence. For these reasons, it is deduced that composites and doping coexist and are present in the ZnO-Nd film.

Author Contributions: M.P.B. and G.G.S. wrote, conceived, and designed the experiments; A.C.S., J.M.G.J. and E.R.A. provided resources and systems; R.R.T., F.G.N.C., C.M.R. and R.G.I. provided support in review and editing this work. All authors have read and agreed to the published version of the manuscript.

Funding: This research received no external funding.

Data Availability Statement: The original contributions presented in the study are included in the article; further inquiries can be directed to the corresponding author.

Acknowledgments: The authors thank CONAHCYT and ICUAP for the use of the facilities.

Conflicts of Interest: The authors declare no conflicts of interest. The funders had no role in the design of the study; in the collection, analyses, or interpretation of data; in the writing of the manuscript; or in the decision to publish the results.

References

1. Sauvik, R.; Ahmaruzzaman, M. ZnO nanostructured materials and their potential applications: Progress, challenges and perspectives. *Nanoscale Adv.* **2022**, *4*, 1868–1925. [CrossRef] [PubMed]
2. Yıldırım, A.K.; Altıokka, B. Effect of Potential on Structural, Morphological and Optical Properties of ZnO Thin Films Obtained by Electrodeposition. *J. Mater. Sci. Eng.* **2015**, *B5*, 107–112. [CrossRef]
3. Chebil, W.; Fouzri, A.; Azeza, B.; Sakly, N.; Mghaieth, R.; Lusson, A.; Sallet, V. Comparison of ZnO thin films on different substrates obtained by sol-gel process and deposited by spin-coating technique. *Indian J. Pure Appl. Phys.* **2015**, *53*, 521–529.
4. Singh, A.K.; Singh, S.K. 7—Optical properties of ZnO. In *Nanostructured Zinc Oxide*; Elsevier: Amsterdam, The Netherlands, 2021; pp. 189–208. [CrossRef]
5. Mursal; Irhamni; Bukhari; Jalil, Z. Structural and Optical Properties of Zinc Oxide (ZnO) based Thin Films Deposited by Sol-Gel Spin Coating Method. *J. Phys. Conf. Ser.* **2018**, *1116*, 032020. [CrossRef]
6. Wisz, G.; Virt, I.; Sagan, P.; Potera, P.; Yavorskyi, R. Structural, Optical and Electrical Properties of Zinc Oxide Layers Produced by Pulsed Laser Deposition Method. *Nanoscale Res. Lett.* **2017**, *12*, 253. [CrossRef] [PubMed]
7. Kang, Y.; Yu, F.; Zhang, L.; Wang, W.; Chen, L.; Li, Y. Review of ZnO-based nanomaterials in gas sensors. *Solid State Ion.* **2021**, *360*, 115544. [CrossRef]
8. Dash, R.; Mahender, C.; Sahoo, P.K.; Soam, A. Preparation of ZnO layer for solar cell application. *Mater. Today Proc.* **2021**, *41*, 161–164. [CrossRef]
9. Galeazzi, R.; Panzo, I.J.; Becerril, T.; Morales, C.; Rosendo, E.; Silva, R.; Trujillo, R.; Coyopol, A.; Caballero, F.G.; Yarce, L. Physicochemical conditions for ZnO films deposited by microwave chemical bath deposition. *RSC Adv.* **2018**, *8*, 8662–8670. [CrossRef]
10. Jangir, L.K.; Kumari, Y.; Kumari, P. 13—Zinc oxide-based light-emitting diodes and lasers. In *Nanostructured Zinc Oxide*; Elsevier: Amsterdam, The Netherlands, 2021; pp. 351–374. [CrossRef]
11. Jafarirad, S.; Salmasi, M.; Divband, B.; Sarabchi, M. Systematic study of Nd^{3+} on structural properties of ZnO nanocomposite for biomedical applications; in-vitro biocompatibility, bioactivity, photoluminescence and antioxidant properties. *J. Rare Earths* **2018**, *37*, 508–514. [CrossRef]
12. Gutiérrez, D.; Salgado, G.; Coyopol, A.; Rosendo, E.; Romano, R.; Morales, C.; Benítez, A.; Severiano, F.; Herrera, A.M.; González, F. Effect of the Deposit Temperature of ZnO Doped with Ni by HFCVD. *Materials* **2023**, *16*, 1526. [CrossRef] [PubMed]
13. Herrera, V.; Becerril, T.; Cervantes, E.; Salgado, G.; Galeazzi, R.; Morales, C.; Rosendo, E.; Coyopol, A.; Romano, R.; Caballero, F.G. Highly Visible Photoluminescence from Ta-Doped Structures of ZnO Films Grown by HFCVD. *Crystals* **2018**, *8*, 395. [CrossRef]
14. Rani, T.D.; Tamilarasan, K.; Elangovan, E.; Leela, S.; Ramamurthi, K.; Thangaraj, K.; Himcinschi, C.; Trenkmann, I.; Schulze, S.; Hietschold, M.; et al. Structural and optical studies on Nd doped ZnO thin films. *Superlattices Microstruct.* **2015**, *77*, 325–332. [CrossRef]

15. Poongodi, G.; Kumar, R.M.; Jayavel, R. Structural, optical and visible light photocatalytic properties of nanocrystalline Nd doped ZnO thin films prepared by spin coating method. *Ceram. Int.* **2015**, *41*, 4169–4175. [CrossRef]
16. Nistor, M.; Millon, E.; Cachoncinlle, C.; Ghica, C.; Hebert, C.; Perrière, J. Nd-doped ZnO films grown on c-cut sapphire by pulsed-electron beam deposition under oblique incidence. *Appl. Surf. Sci.* **2021**, *563*, 150287. [CrossRef]
17. Yatskiv, R.; Grym, J.; Bašinová, N.; Kučerová, Š.; Vaniš, J.; Piliai, L.; Vorokhta, M.; Veselý, J.; Maixner, J. Defect-mediated energy transfer in ZnO thin films doped with rare-earth ions. *J. Lumin.* **2023**, *253*, 119462. [CrossRef]
18. López, R.; Díaz, T.; García, G.; Galeazzi, R.; Rosendo, E.; Coyopol, A.; Pacio, M.; Juárez, H.; Oliva, A.I. Structural Properties of Zn-ZnO Core-Shell Microspheres Grown by Hot-Filament CVD Technique. *J. Nanomater.* **2012**, *2012*, 7. [CrossRef]
19. López, R.; Díaz, T.; Rosendo, E.; García, G.; Coyopol, A.; Juárez, H. Propiedades fotoluminiscentes de películas zno: A-Sio_x obtenidas por la técnica cvd asistido por filamento caliente. *Rev. Latinoam. Metal. Mater.* **2011**, *31*, 59–63.
20. López, R.; Díaz, T.; García, G.; Rosendo, E.; Galeazzi, R.; Juárez, H. Caracterización estructural y óptica de compósitos ZnO-SiOx obtenidos por la técnica Cat-CVD. *Superf. Vacío* **2011**, *24*, 76–80.
21. López, R.; García, G.; Díaz, T.; Coyopol, A.; Rosendo, E.; Galeazzi, R.; Juárez, H.; Pacio, M. Low temperature growth of Zn-ZnO microspheres by atomic hydrogen assisted-HFCVD. *IOP Conf. Ser. Mater. Sci. Eng.* **2013**, *45*, 012016. [CrossRef]
22. López, R.; Díaz, T.; García, G.; Rosendo, E.; Galeazzi, R.; Coyopol, A.; Juárez, H.; Pacio, M.; Morales, F.; Oliva, A.I. Fast Formation of Surface Oxidized Zn Nanorods and Urchin-Like Microclusters. *Adv. Mater. Sci. Eng.* **2014**, *2014*, 4. [CrossRef]
23. López, R.; García, G.; Coyopol, A.; Díaz, T.; Rosendo, E. Effect of nitrogen gas in the agglomeration and photoluminescence of Zn-ZnO nanowires after high-temperature annealing. *Rev. Mex. Física* **2016**, *62*, 1–4.
24. Chauhan, S.; Kumar, M.; Chhoker, S.; Katyal, S. Structural, vibrational, optical and magnetic properties of sol–gel derived Nd doped ZnO nanoparticles. *J. Mater. Sci. Mater. Electron.* **2013**, *24*, 5102–5110. [CrossRef]
25. Mousavi, M.; Zinatloo, S.; Ghodrati, M. One-step sonochemical synthesis of $Zn(OH)_2/ZnV_3O_8$ nanostructures as a potent material in electrochemical hydrogen storage. *J. Mater. Sci. Mater. Electron.* **2020**, *31*, 17332–17338. [CrossRef]
26. Arunachalam, S.; Kirubasankar, B.; Murugadoss, V.; Vellasamy, D.; Angaiah, S. Facile synthesis of electrostatically anchored $Nd(OH)_3$ nanorods onto graphene nanosheets as high capacitance electrode material for supercapacitors. *New J. Chem.* **2018**, *42*, 2923–2932. [CrossRef]
27. Yuan, L.; Wang, C.; Cai, R.; Wang, Y.; Zhou, G. Temperature- dependent growth mechanism and microstructure of ZnO nanostructures grown from the thermal oxidation of zinc. *J. Cryst. Growth* **2014**, *390*, 101–108. [CrossRef]
28. Olvera, D.; Becerril, T.; Salgado, G.; Solís, A.; Andrés, E.; Isasmendi, R.; Sierra, R.; Ruiz, C.; Trujillo, R.; Caballero, F.G. Photoluminiscent Enhancement by Effect of Incorporation Nickel in ZnO Films Grown. *Eur. J. Eng. Technol. Res.* **2021**, *6*, 177–180. [CrossRef]
29. Honglin, L.; Yingbo, L.; Jinzhu, L.; Ke, Y. Experimental and first-principles studies of structural and optical properties of rare earth (RE=La, Er, Nd) doped ZnO. *J. Alloys Compd.* **2014**, *617*, 102–107. [CrossRef]
30. Kumar, S.; Sahare, P.D. Nd-doped ZnO as a multifunctional nanomaterial. *J. Rare Earths* **2012**, *30*, 761. [CrossRef]
31. Jayachandraiah, C.; Divya, A.; Kumar, K.; Krishnaiah, G. Effect of Nd on Structural and Optical Properties of Nd Doped ZnO Nanoparticles. In Proceedings of the International Conference on Advanced Nanomaterials & Emerging Engineering Technologies, Chennai, India, 24–26 July 2013.
32. Zhao, Z.; Song, J.; Zheng, J.; Lian, J. Optical properties and photocatalytic activity of Nd-doped ZnO powders. *Trans. Nonferrous Met. Soc. China* **2014**, *24*, 1434–1439. [CrossRef]
33. Wang, M.; Jiang, L.; Jung, E.; Hong, S. Electronic structure and optical properties of $Zn(OH)_2$: LDA+U calculations and intense yellow luminescence. *RSC Adv.* **2015**, *5*, 87496–87503. [CrossRef]

crystals

MDPI

Article

The Influence of Variations in Synthesis Conditions on the Phase Composition, Strength and Shielding Characteristics of $CuBi_2O_4$ Films

Dauren B. Kadyrzhanov [1], Medet T. Idinov [2], Dmitriy I. Shlimas [1,3,*] and Artem L. Kozlovskiy [1,3]

1 Engineering Profile Laboratory, L.N. Gumilyov Eurasian National University, Satpayev St., Astana 010008, Kazakhstan; kadyrzhanov.d@enu.edu.kz (D.B.K.); kozlovskiy.a@inp.kz (A.L.K.)
2 Department of Technical Physics and Thermal Power Engineering, NJSC Shakarim University of Semey, Shygayev St., Semipalatinsk 071400, Kazakhstan; medet@nucmed.kz
3 Laboratory of Solid State Physics, The Institute of Nuclear Physics, Ibragimov St., Almaty 050032, Kazakhstan
* Correspondence: shlimas@inp.kz

Abstract: This paper presents the results of the influence of variation of the synthesis conditions of $CuBi/CuBi_2O_4$ films with a change in the applied potential difference, as well as a change in electrolyte solutions (in the case of adding cobalt or nickel sulfates to the electrolyte solution) on changes in the phase composition, structural parameters and strength characteristics of films obtained using the electrochemical deposition method. During the experiments, it was found that, in the case of the addition of cobalt or nickel to the electrolyte solutions, the formation of films with a spinel-type tetragonal $CuBi_2O_4$ phase is observed. In this case, a growth in the applied potential difference leads to the substitution of copper with cobalt (nickel), which in turn leads to an increase in the structural ordering degree. It should be noted that, during the formation of $CuBi/CuBi_2O_4$ films from solution–electrolyte №1, the formation of the $CuBi_2O_4$ phase is observed only with an applied potential difference of 4.0 V, while the addition of cobalt or nickel sulfates to the electrolyte solution results in the formation of the tetragonal $CuBi_2O_4$ phase over the entire range of the applied potential difference (from 2.0 to 4.0 V). Studies have been carried out on the strength and tribological characteristics of synthesized films depending on the conditions of their production. It has been established that the addition of cobalt or nickel sulfates to electrolyte solutions leads to an increase in the strength of the resulting films from 20 to 80%, depending on the production conditions (with variations in the applied potential difference). During the studies, it was established that substitution of copper with cobalt or nickel in the composition of $CuBi_2O_4$ films results in a rise in the shielding efficiency of low-energy gamma radiation by 3.0–4.0 times in comparison with copper films, and 1.5–2.0 times for high-energy gamma rays, in which case the decrease in efficiency is due to differences in the mechanisms of interaction of gamma quanta, as well as the occurrence of secondary radiation as a result of the formation of electron–positron pairs and the Compton effect.

Keywords: radiation shielding; protective materials; thin films; electrochemical synthesis; substitution effect

Citation: Kadyrzhanov, D.B.; Idinov, M.T.; Shlimas, D.I.; Kozlovskiy, A.L. The Influence of Variations in Synthesis Conditions on the Phase Composition, Strength and Shielding Characteristics of $CuBi_2O_4$ Films. *Crystals* **2024**, *14*, 453. https://doi.org/10.3390/cryst14050453

Academic Editor: Sven L. M. Schroeder

Received: 24 March 2024
Revised: 29 April 2024
Accepted: 9 May 2024
Published: 10 May 2024

1. Introduction

Today, the problem of protection from the negative effects of ionizing radiation on living organisms, as well as microelectronics (operating in conditions of increased background radiation), is one of the most important in the modern world [1–3]. There are many reasons for this; first of all, the increased use of various sources of ionizing radiation (non-natural origin) in various industries (energy, medicine, etc.) leads to an increase in the likelihood of exposure to radiation on living organisms, which can lead to negative consequences, including mutations, diseases and deterioration of health [4,5]. At the same time, artificial sources of ionizing radiation are increasingly used in medicine (X-ray machines, gamma knives, proton and heavy ion accelerators for the treatment of tumors), which requires increasing the level of protection from exposure to ionizing radiation for personnel who are directly involved in their maintenance and operation, since classical traditional methods of protection in these cases cannot always be used [6,7].

An important role in developments to shield ionizing radiation is also played by work related to the disposal of nuclear waste at nuclear power plants or nuclear test sites, where it is necessary to comply with strict control and safety standards for long-term storage, as well as to minimize the negative impact of ionizing radiation on the environment [8,9]. Among the most effective materials for these purposes are carbide and nitride ceramics, which have high strength and wear resistance, as well as resistance to radiation damage during prolonged exposure and the accumulation of high doses of damage [10–12].

Recently, much attention has been paid to research on the prospects of using thin films or thin-film coatings as shielding protective materials, interest in which is primarily due to the possibility of using them to create local protection of key components of microelectronic devices operating under conditions of exposure to increased background radiation (in spacecraft, satellites or nuclear power plants) [13–15]. Thus, the use of $CuBi/CuBi_2O_4$ films as promising materials for protection against the negative effects of electromagnetic radiation was shown in [16,17]. Much attention is paid to composite materials and films consisting of oxide compounds and polymer films as protective shielding materials [18,19]. Interest in such research is due to the possibility of expanding the classes of protective materials, as well as thin, lightweight and sufficiently flexible shields that can be used to protect complex-profile objects. Moreover, these technological solutions are based on technologies for combining light and heavy elements, as well as binding polymer matrices, which serve both as a basis for applied coatings and as matrices in which oxide particles are equally distributed throughout the entire volume [20]. Therefore, for example, in [21], the authors show the prospects for using composite materials based on polymer matrices with oxide nanoparticles placed in them as shielding materials with high shielding efficiency, which are due to the presence of oxides such as WO_3 and Bi_2O_3. Also, the use of composite materials in the form of coatings or films is aimed primarily at reducing the cost of production of microelectronic devices for operation in flows of ionizing radiation, for protection against which the classic scheme of duplicating key components is used in order to avoid failures due to radiation damage caused by exposure to ionizing radiation [22,23]. At the same time, the use of these technological solutions makes it possible to solve the issue of protecting key components from the negative effects of ionizing radiation, as well as to reduce the load on the weight and dimensions of microcircuits, which plays a very important role in the case of spacecraft, since transporting each kilogram into orbit requires large amounts of fuel. Thin films play an important role in shielding not only various types of ionizing radiation (gamma, neutron or electron), but also in protecting against electromagnetic influence, which can lead to failures in microelectronic devices along with radiation damage [24,25].

Based on the analysis of the main areas of research in the field of development of shielding materials and their application, the goal of this research was formulated, which is aimed at creating effective shielding materials that can have a positive effect in this direction. The key purpose of this work is to develop a technology for producing composite films based on compounds of copper, bismuth, nickel, cobalt and their oxide compounds such as $CuBi/CuBi_2O_4$, as well as to evaluate their use as shielding materials for protection against the negative effects of ionizing radiation [26,27].

The novelty and relevance of this research is based on the possibility of obtaining new types of high-strength shielding materials based on $CuBi/CuBi_2O_4$ with the partial replacement of copper with cobalt or nickel, which can be used to protect against the negative effects of ionizing radiation (gamma and X-rays), and also be used as flexible protective shields for shielding, which have good corrosion resistance and wear resistance (resistance to mechanical stress). The choice as research objects for the development of protective shielding films based on $CuBi/CuBi_2O_4$ is due to the high density of these compounds (about 8.5–8.6 g/cm^3), comparable to alternative ceramic or glass-like materials based on oxide compounds of tellurium, tungsten, bismuth and lead (the density of these structures varies from 5.0 to 10 g/cm^3, depending on the number of components in the material and their stoichiometric composition), as well as the possibility of creating protective films with great flexibility, which is due to the polymer matrix being used as a substrate for the synthesized films, which allows them to be used in shielding devices of complex geometry. Moreover, consideration of the possibility of partial replacement of copper in the $CuBi_2O_4$ composition with related elements such as nickel, cobalt or iron is due to the possibility of increasing not only the shielding efficiency due to changes in the charge number (Z_{eff}) when replacing copper, but also increasing the strength parameters (hardness, wear resistance, corrosion resistance) due to the formation of a more stable film structure. The choice of nickel and cobalt as components to replace copper is due to their similarity of atomic radii (r_{Cu}—128 pm, r_{Co}—125 pm, r_{Ni}—124 pm), as well as their electrode reduction potentials, which makes it possible to use the method of varying the applied potential difference to vary the ratio of elements in the composition of the films. The use of nickel and cobalt in electrochemical deposition, meanwhile, has proven itself rather well in order to obtain sufficiently strong and wear-resistant coatings that have higher corrosion resistance rates than copper coatings, which have a tendency for rapid oxidation.

2. Materials and Research Methods

2.1. Preparation of Solutions–Electrolytes for the Synthesis of Thin Films

The following components were used as initial components for preparing electrolyte solutions: solution–electrolyte №1 to obtain CuBi/$CuBi_2O_4$ films–$CuSO_4{\cdot}5H_2O$ (238 g/L), $Bi_2(SO_4)_3$ (10 g/L), H_2SO_4 (21 g/L); solution–electrolyte №2 to obtain CuBi/$CuBi_2O_4$ films with partial substitution of copper by nickel–$CuSO_4{\cdot}5H_2O$ (200 g/L), $Co_2SO_4{\cdot}7H_2O$ (40 g/L), $Bi_2(SO_4)_3$ (10 g/L), H_2SO_4 (21 g/L); solution–electrolyte №3 to obtain CuBi/$CuBi_2O_4$ films with partial substitution of copper with cobalt–$CuSO_4{\cdot}5H_2O$ (200 g/L), $Ni_2SO4{\cdot}7H_2O$ (40 g/L), $Bi_2(SO_4)_3$ (10 g/L), H_2SO_4 (21 g/L). Solutions–electrolytes were prepared by dissolving all components in given proportions in distilled water, mixing the components using magnetic stirrers at a constant stirring speed (50–100 rpm) and a temperature of 45–50 °C, in order to achieve complete dissolution of all the salts used. The selection of the ratio of components for the preparation of electrolyte solutions was carried out experimentally, the purpose of which was to test the modes of film production and the ability to control the stoichiometric composition of the resulting films. After stirring in order to achieve complete dissolution of all salts, the resulting solutions were kept for two to three hours until they reached room temperature, in order to prevent the influence of temperature factors on the processes of electrochemical reduction in the metal deposit in the form of films on the surface of the substrates.

For the synthesis of films, a PLA cell printed using additive technologies was chosen. The distance between the electrodes was fixed and amounted to 3 cm; for this, the upper electrode was placed on a special platform located at a fixed distance from the lower electrode. Chlorine, a silver electrode, was chosen as a reference electrode, with the help of which the deposition parameters were monitored. The deposition time of the films was fixed and amounted to 20 min. At the same time, it was experimentally established that an increase in deposition time of more than 60 min leads to a decrease in the deposition rate (a decrease in current density was observed), which is due to the depletion of the electrolyte solution over time.

The main method for producing thin films was the electrochemical synthesis method, which is based on the reduction in a metal deposit from aqueous solutions–electrolytes on the surface of the cathode when an electric current passes through the solution–electrolyte. Copper plates of the same area were used as the anode and cathode, varying which makes it possible to obtain films of different sizes. To study the effect of variation in synthesis conditions on the phase composition of the films, the variation range of the applied potential difference was chosen from 2.0 to 4.0 V with a step of 0.5 V, which made it possible to obtain films with a controlled phase composition, as well as a different ratio of elements, the change in which occurs with varying synthesis conditions. In the case of using solution–electrolyte №1 (without additives), an elevation in the applied potential difference above 4.0 V leads to structural ordering and dominance of the $CuBi_2O_4$ phase over the Cu(Bi) phase (see the example X-ray diffraction patterns of samples obtained using solution–electrolyte №1 in Figure 1). If the applied potential difference increases above 4.0 V when using electrolyte solutions with the addition of cobalt sulfates (solution №2) and nickel sulfates (solution №3), a rapid release of oxygen is observed in the electrochemical cell, which leads to uneven deposition associated with partial overlap of the cathode surface. Based on this, the range of applied potential differences in this experiment was chosen from 2.0 to 4.0 V.

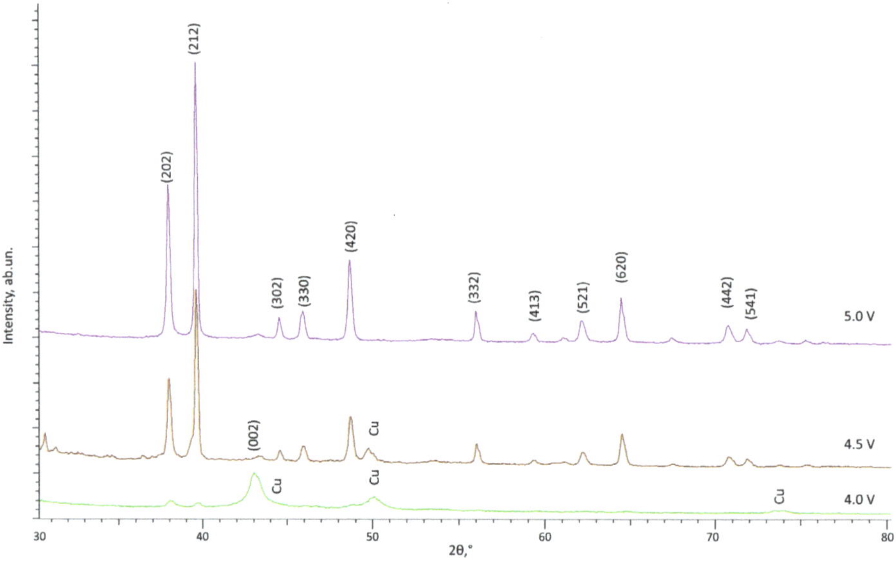

Figure 1. Results of a comparative analysis of X-ray diffraction patterns of $Cu(Bi)/CuBi_2O_4$ films obtained at applied potential differences in the range from 4.0 to 5.0 V.

2.2. *Characterization of the Films under Study*

To determine the surface morphology of the resulting thin films depending on variations in synthesis conditions (changes in electrolyte solution, difference in applied potentials), the atomic force microscopy method was used, implemented on a Smart SPM microscope (AIST-NT, Zelenograd, Russia) in semi-contact shooting mode. Based on the data obtained, 3D images of the sample surface were constructed, reflecting changes in surface morphology during film growth (in the case of changes in deposition time), as well as with variations in synthesis conditions (changes in the electrolyte solution and the difference in applied potentials).

Determination of the phase composition of the synthesized films depending on the synthesis conditions (with variations in the applied potential difference, as well as changes in electrolyte solutions) was carried out using the X-ray diffraction method. The recording of X-ray diffraction patterns was carried out on a D8 ADVANCE ECO X-ray diffractometer (Bruker, Karlsruhe, Germany) in the Bragg–Brentano geometry (2θ = 30–100°, with a step of 0.03° and a diffraction pattern acquisition time at a point of 1 s). The determination of the phase composition was carried out by comparative analysis of the position of diffraction lines on the obtained experimental diffraction patterns with card values from the PDF-2 (2016) database, considering possible distortions of the structure (shift in diffraction lines) caused by the deposition process. At the same time, comparison of the obtained data with the experimental data was carried out when the card (reference) values from the PDF-2 database coincided with the experimental data with an accuracy of about 90%.

Determination of the elemental composition of the film samples under study, depending on the production conditions, was carried out by recording energy-dispersive spectra on a TM3030 scanning electron microscope (SEM) (Hitachi, Tokyo, Japan), equipped with an attachment for energy-dispersive analysis. The accuracy of the measurements was achieved by taking about 10–15 spectra from different areas of the samples, as well as subsequent assessment of the distribution of elements in the composition of the films using the mapping method (determining the equiprobable distribution of elements over large areas of the samples).

Measurements of the optical properties of the synthesized films depending on the production conditions were performed on a SPECORD 250 PLUS UV-Vis spectrophotometer (Analytik Jena, Jena, Germany). The measurements were carried out in the wavelength range from 300 to 1000 nm with a step of 1 nm.

Determination of the strength characteristics of the synthesized films depending on the conditions of their preparation was carried out using the indentation method, implemented using a Duroline M1 microhardness tester (Metkon, Bursa, Turkey). A Vickers diamond pyramid was used as an indenter; the load on the indenter was 10 N, which made it possible to measure the hardness of thin films without the influence of the indenter on the substrate.

Tests to measure the dry friction coefficient, as well as determine the effectiveness of the effect of replacing copper with cobalt or nickel in the composition of films to external mechanical influences, were performed using a UNITEST 750 tribometer (Ducom Instruments, Bengaluru, India). The tests were carried out by successive tests using a ball-shaped indenter, which was applied to the surface under a load of 100 N. The number of friction repetition cycles was 20,000. Based on the obtained tribological test data, the dry friction coefficient (as well as the dynamics of its change depending on the number of cycles of successive tests), as well as the wear profile of coatings, indicating degradation of the film surface depending on their type, were determined.

2.3. Determination of Shielding Efficiency

The assessment of the shielding ability of $CuBi/CuBi_2O_4$ films in gamma radiation shielding was carried out using the classical scheme of shielding experiments [28,29]. The efficiency of shielding and intensity reduction in gamma radiation was assessed using a standard method for assessing the intensity of recorded gamma radiation with a certain energy at a distance of 10 cm from the source of gamma rays using a NaI detector. Co^{57} (130 keV), Cs^{137} (660 keV) and Na^{22} (1230 keV) were used as sources of gamma quanta, which made it possible to simulate the processes of interaction of gamma rays with matter, including the photoelectric effect, the Compton effect as well as the formation of electron–positron pairs. The shielding efficiency was determined by changes in the spectra of recorded gamma rays before and after shielding. In the case of using thin shielding films, the quality of the spectrum deteriorates sharply, due to a decrease in statistics, as well as a decrease in line intensity. An increase in the statistical spread of points on the spectrum indicates a small number of recorded effects associated with the passage and subsequent interaction of gamma quanta. The shielding efficiency was determined by comparing the intensity values recorded without a protective shield and using protective shields made of synthesized films.

3. Results and Discussion

3.1. The Influence of Variations in Solutions–Electrolytes on Changes in the Phase Composition and Structural Parameters of Synthesized $CuBi/CuBi_2O_4$ Films

One of the ways to vary the phase composition of films obtained using the electrochemical deposition method is to change the synthesis conditions (variation of the applied potential difference). As is known, a rise in the applied potential difference in the case of two or three-component electrolyte solutions leads to a change in the rate of reduction in metal ions in the solution, which in turn leads to a change in the elemental composition of the resulting structures, and as a consequence the possibility of changing the phase composition due to the structural formation of films from various elements. In this case, the dominance of the reduction rate (reduction potential) of metal ions from aqueous solutions–electrolytes can lead to the dominance of one of the elements in the resulting structures which, under certain conditions or concentrations, can lead to the formation of new structural elements in the form of inclusions of new phases, or a complete phase transformation of the resulting films. The most reliable method for determining phase and structural changes in films is the X-ray diffraction method, the use of which makes it possible to obtain data not only on phase changes in films when varying the conditions for their production, but also to determine the influence of variations in the applied potential difference (as a consequence, changes in the rate of reduction in metal ions) on the structural parameters of the resulting films and the structural ordering degree.

Figure 2a shows X-ray diffraction patterns of the studied samples of $CuBi/CuBi_2O_4$ films obtained from solution–electrolyte №1, with variation in the applied potential difference, the changing of which makes it possible to increase the deposition rate by changing the rate of reduction in metal deposits from sulfuric acid aqueous solutions.

According to the assessment of the general appearance of the presented X-ray diffraction patterns of the studied samples, it can be concluded that a change in the applied potential difference leads to two types of structural changes: (1) a change in the structural ordering degree, expressed in a change in the shape of the diffraction maxima, as well as a change in their intensity (texturing effect); (2) a change in the phase composition of the films, which manifests itself due to the appearance of new diffraction reflections in diffraction patterns at high potential differences. At the same time, the ratio of the intensities of diffraction reflections and background radiation indicates the polycrystalline structure of the resulting films, as well as a fairly high degree of structural ordering (crystallinity degree) of the films, a change in which is observed with variations in synthesis conditions (changes in the applied potential difference).

The general appearance of the presented diffraction patterns, depending on changes in synthesis conditions (with variations in the difference in applied potentials), indicates not only changes in the structural ordering degree (expressed in changes in the shape and intensity of diffraction reflections), but also the processes of phase transformations that appear at applied potential differences above 3.0 V. In the case of applied potential differences from 2.0 to 3.0 V, the main positions of the diffraction reflections presented in the X-ray diffraction patterns correspond to the cubic phase of Cu (PDF-00-004-0836), the formation of which is due to the processes of electrochemical reduction in the metal deposit, as well as the potential for the reduction in copper from sulfuric acid aqueous solutions–electrolytes [30,31]. In this case, the shape and angular position of the diffraction reflections indicate a deformation distortion of the crystal lattice of the tensile type (shift of reflections to the region of small angles), which can be explained by the effect of partial substitution of copper ions by bismuth ions at the crystal lattice sites, the ionic radius of which (1.2 Å) is significantly larger than the ionic radius of copper (0.98 Å). In this case, the shift in the position of the diffraction reflections relative to the initial position (determined for samples of $CuBi/CuBi_2O_4$ films obtained at a potential difference of 2.0 V) can be explained by an increase in the bismuth content in the films, which is observed according to energy dispersion analysis data (see results presented in Figure 2a). Also, changes in crystal lattice parameters (their increase) are evidenced by the data presented in Table 1, which were determined using the Nelson–Taylor technique, used to estimate structural parameters by selecting a certain number of approximating functions when analyzing the shape and position of diffraction reflections [32]. At the same time, in the case of an increase in the applied potential difference from 2.0 to 3.5 V, not only a change in the shape of the main diffraction reflections is observed, indicating the structural ordering degree, but also the appearance of a texture effect, which is most pronounced for film samples obtained at potential differences of 3.0–3.5 V, for which an increase in the intensity of the diffraction reflection is observed at the angular position of 2θ = 74.0–74.5°, comparable in magnitude to the intensity of the diffraction reflection at 2θ = 43.0–43.5°. This change in the intensities of diffraction reflections with increasing difference in applied potentials is due to the fact that the formation of grains in the film structure occurs along two selected textural directions, which indicates the occurrence of the effect of texture misorientation of grains, which manifests itself for nanostructured materials obtained by electrochemical deposition [33,34].

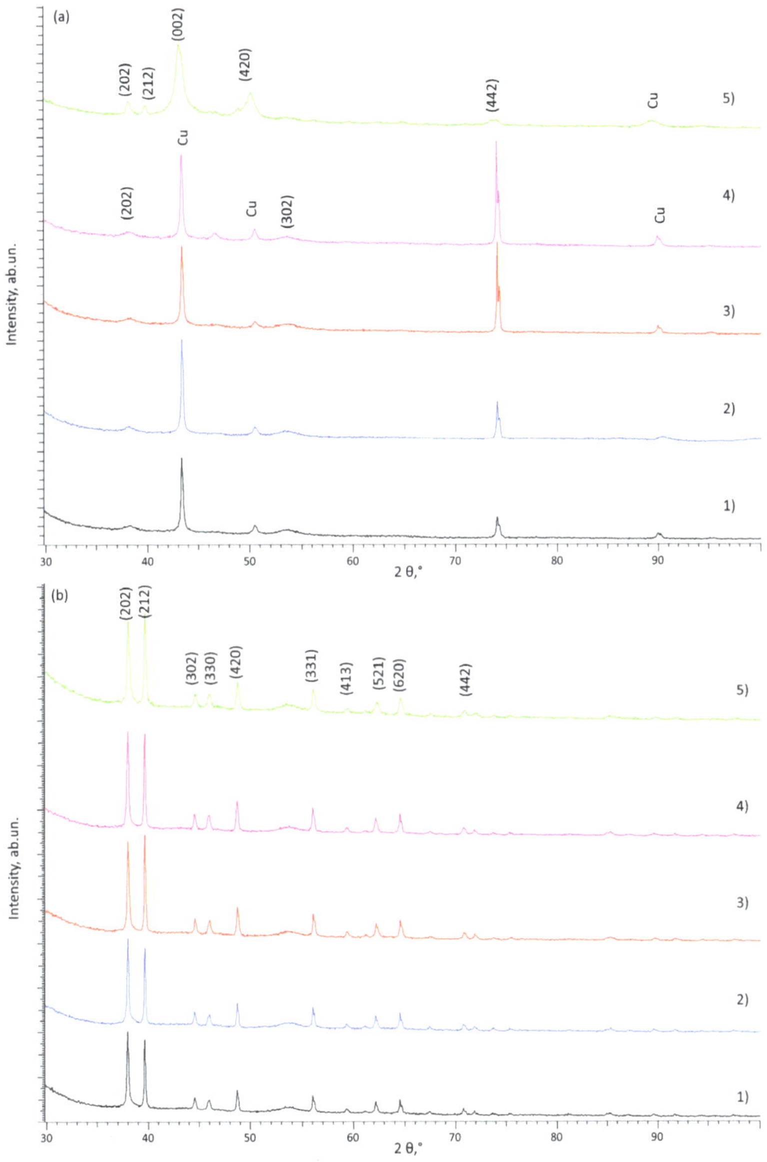

Figure 2. *Cont.*

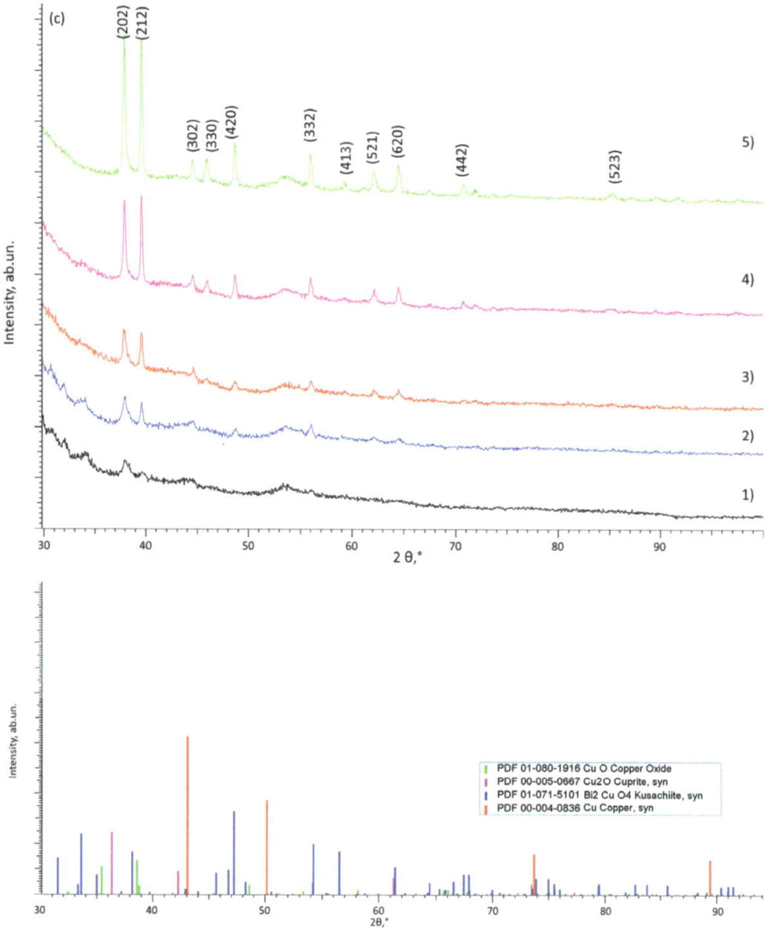

Figure 2. Results of X-ray diffraction of the studied samples of films obtained using the electrochemical deposition method with variation in the applied potential difference: (1) = 2.0 V, (2) = 2.5 V, (3) = 3.0 V, (4) = 3.5 V, and (5) = 4 V: (**a**) when using the electrolyte composition to obtain CuBi/$CuBi_2O_4$ films; (**b**) when using the electrolyte composition to obtain Cu(Co)Bi_2O_4 films, (**c**) when using the electrolyte composition to obtain Cu(Ni)Bi_2O_4 films.

Figure 3 demonstrates the mapping results of the film samples under study depending on the type of electrolyte solution used, which reflects the isotropic distribution of elements in the film composition over the surface. The figure also reveals data on the morphological features of the synthesized films, found using the scanning electron microscopy method The overall appearance of the data presented on the element distribution maps indicates the isotropy of the distribution of elements in the composition of the films. In this case, an alteration in synthesis conditions, in particular, variations in the applied potential difference, leads to the displacement of copper and an increase in the composition of bismuth and oxygen, in the case of using electrolyte solution №1, which has good agreement with the data of X-ray phase analysis. When cobalt or nickel sulfates are added to the electrolyte solution, according to the data presented in Figure 3b,c, it is clear that a growth in the applied potential difference leads to a rise in the content of cobalt or nickel in the composition of the films, while their uniform distribution over the volume is observed indicating a partial replacement of copper in the composition.

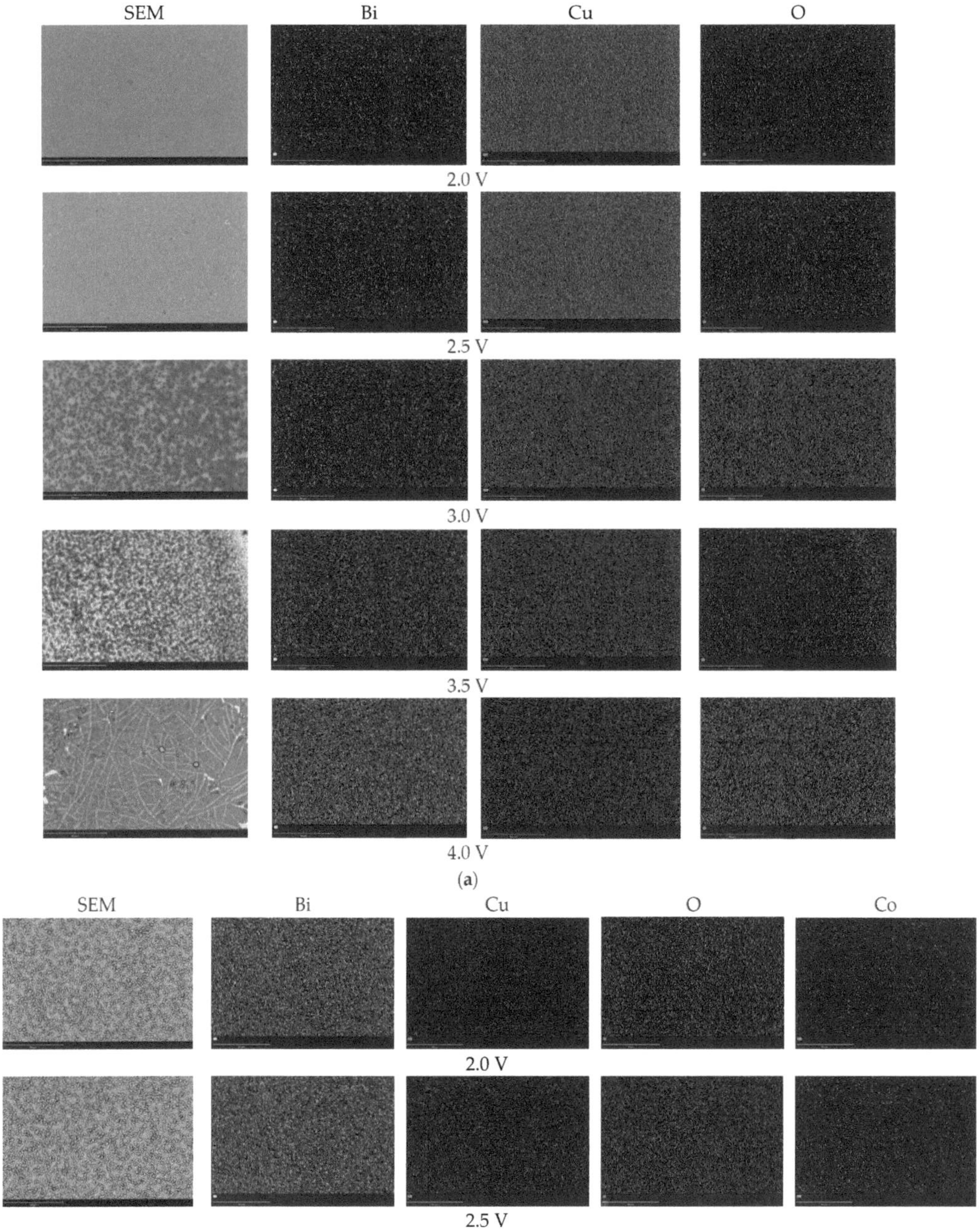

Figure 3. *Cont.*

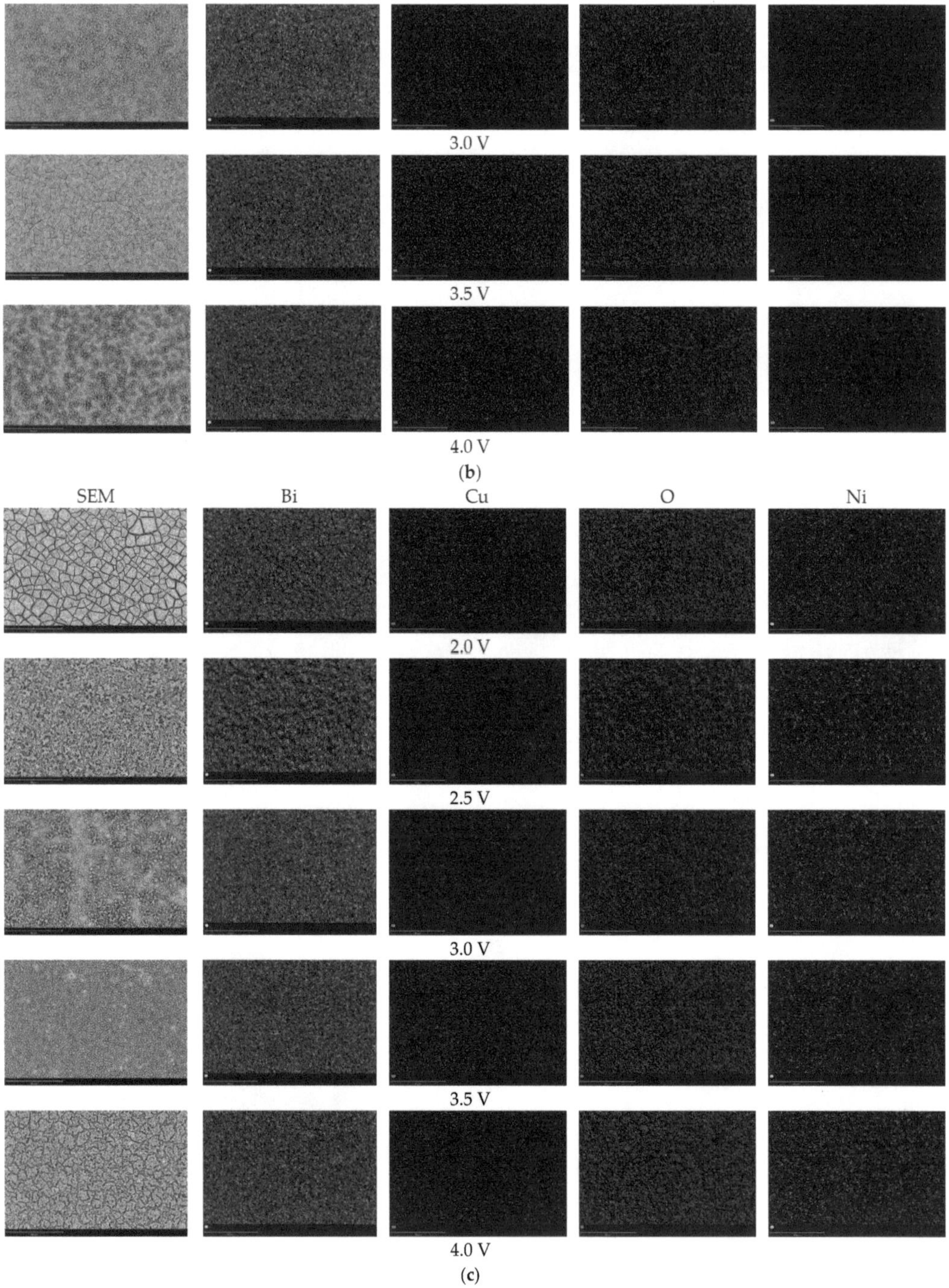

Figure 3. Mapping results reflecting the isotropy of the distribution of elements in the composition of films synthesized under various conditions: (**a**) when using the electrolyte composition to obtain CuBi/$CuBi_2O_4$ films; (**b**) when using the electrolyte composition to obtain Cu(Co)Bi_2O_4 films; (**c**) when using the electrolyte composition to obtain Cu(Ni)Bi_2O_4 films (under the diffraction patterns are line diagrams of cards from the PDF-2 (2016) database).

By analyzing changes in the elemental composition of the resulting CuBi/$CuBi_2O_4$ films with changes in the applied potential difference, we can draw the following conclusions (see data in Figure 4) associated with the fact that the main changes in the elemental composition of films with an increase in the applied potential difference from 2.0 to 2.5 V occur due to an increase in the weight contribution of bismuth, the content of which increases from 6.6 at.% to 12.5 at.%. Moreover, such a change is in good agreement with the data on changes in the parameters of the crystal lattice presented in Table 1, as well as with the assumption made about the partial substitution of copper ions by bismuth ions during the formation of films. At the same time, the absence in the presented X-ray diffraction patterns of reflections characteristic of bismuth or other compounds of the substitutional solid solution type can be explained by low concentrations of bismuth in the structure of the films, as well as a well-structured crystal lattice of copper, the reflections of which are quite clearly visible in the diffraction patterns. It should also be noted that, when the applied potential difference increases to 3.0 V, a low oxygen content (no more than 2.5%) is observed in the structure of the resulting films, the presence of which can be explained by the structural formation of the films which, in the case of high potential differences, is accompanied by a rapid release of oxygen, which can penetrate into the films being formed, filling vacancies or interstices. In this case, also, the observed increase in the bismuth content in the films is in good agreement with the data on changes in structural parameters (data on the crystal lattice parameters are presented in Table 1 for the sample obtained at a potential difference of 3.0 V).

In the case of samples obtained at potential differences of 3.5 V, an increase in the content of bismuth (about 21 at.%) and oxygen (more than 5 at.%) is observed, which is consistent with X-ray diffraction data, according to which the diffraction pattern of the sample under study shows the appearance of low-intensity reflections at angular positions 2θ = 38.0, 40.5 and 46.0°, characteristic of the tetragonal phase of $CuBi_2O_4$ (PDF-01-071-5101); however, it is impossible to establish the weight contribution of it due to its low intensity. At the same time, for samples obtained at a potential difference of 4.0 V, the position of diffraction reflections is characteristic of the tetragonal phase of $CuBi_2O_4$, and when analyzing the elemental composition data, it was found that, at a given potential difference, a high content of oxygen and bismuth is observed in the structure of the films.

Table 1. Crystal lattice parameters of the films under study in the case of using different solutions–electrolytes.

Phase	When Using Solution–Electrolyte №1				
	Applied Potential Difference, V				
	2.0	2.5	3.0	3.5	4.0
Cu–Cubic (PDF-00-004-0836)	a = 3.6086 ± 0.0014 Å *	a = 3.6114 ± 0.0021 Å	a = 3.6148 ± 0.0017 Å	a = 3.6163 ± 0.0022 Å	-
$CuBi_2O_4$–tetragonal (PDF-01-071-5101)	-	-	-	-	a = 8.4607 ± 0.0026 Å, c = 5.8022 ± 0.0025 Å
Phase	When Using Solution–Electrolyte №2				
	Applied Potential Difference, V				
	2.0	2.5	3.0	3.5	4.0
$CuBi_2O_4$–tetragonal (PDF-01-071-5101)	a = 8.4292 ± 0.0023 Å, c = 5.7072 ± 0.0021 Å	a = 8.4359 ± 0.0016 Å, c = 5.6926 ± 0.0023 Å	a = 8.4325 ± 0.0015 Å, c = 5.6758 ± 0.0024 Å	a = 8.4258 ± 0.0023 Å, c = 5.6713 ± 0.0022 Å	a = 8.4191 ± 0.0017 Å, c = 5.6668 ± 0.0022 Å
Phase	When Using Solution–Electrolyte №3				
	Applied Potential Difference, V				
	2.0	2.5	3.0	3.5	4.0
$CuBi_2O_4$–tetragonal (PDF-01-071-5101)	a = 8.4738 ± 0.0024 Å, c = 5.8226 ± 0.0022 Å	a = 8.4622 ± 0.0026 Å, c = 5.8124 ± 0.0023 Å	a = 8.4572 ± 0.0023 Å, c = 5.8069 ± 0.0015 Å	a = 8.4456 ± 0.0024 Å, c = 5.7965 ± 0.0018 Å	a = 8.4373 ± 0.0022 Å, c = 5.7908 ± 0.0014 Å

* parameter refinement was carried out by application of a method based on comparing the position of experimentally obtained diffraction patterns with the data of reference values taken from the PDF-2 (2016) database.

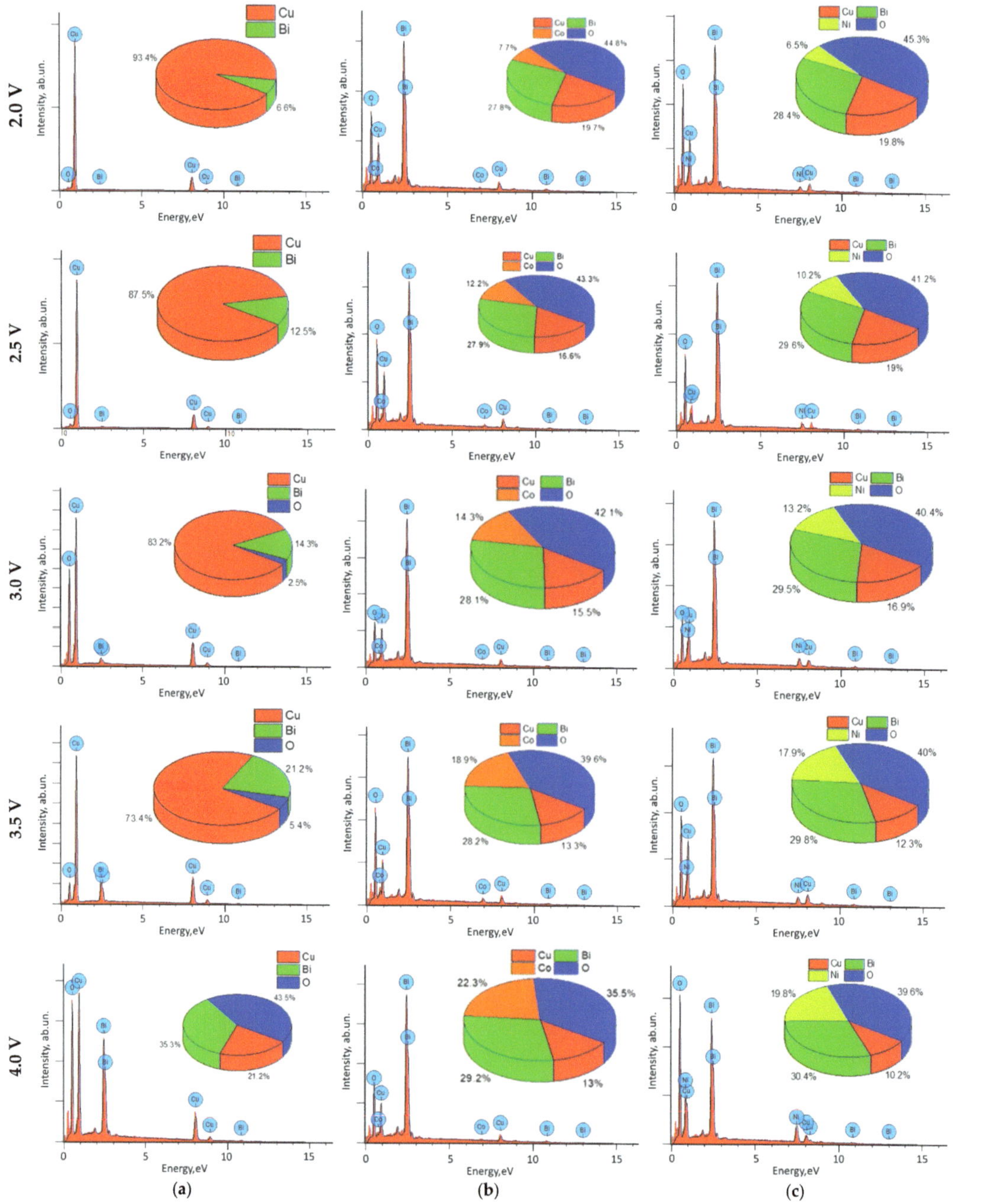

Figure 4. Results of energy dispersive analysis of films: (**a**) when using the electrolyte composition to obtain $CuBi/CuBi_2O_4$ films; (**b**) when using the electrolyte composition to obtain $Cu(Co)Bi_2O_4$ films; (**c**) when using the electrolyte composition to obtain $Cu(Ni)Bi_2O_4$ films.

According to the obtained X-ray phase analysis data, it was established that, when the applied potential difference increases above 3.5 V, a phase transformation of the $Cu(Bi) \rightarrow CuBi_2O_4$ type is observed, which leads to the formation of films with a spinel type of crystal structure and high density. It was established that, in the range of potential differences of 2.0–3.5 V, the dominant phase is the cubic phase of copper, the change in the parameters of the crystal lattice of which indicates the partial substitution of copper ions by bismuth ions while maintaining the cubic type of the crystal lattice. However, when the bismuth content in the films is more than 20 at.%, it leads to the initialization in the structure of the resulting films of phase transformations associated with the formation of the tetragonal $CuBi_2O_4$ phase, which has a spinel type of crystal lattice.

Figure 2b demonstrates the results of X-ray diffraction of samples of $CuBi/CuBi_2O_4$ films when a cobalt sulfate electrolyte was added to the solution, obtained by varying the applied potential difference. The obtained diffraction patterns indicate the polycrystalline structure of the synthesized films, and the observed changes depending on variations in synthesis conditions (changes in the applied potential difference) are characterized by changes in the degree of structural ordering, the change of which is associated with the processes of film formation during electrochemical synthesis.

According to the assessment of the phase composition for the studied $CuBi/CuBi_2O_4$ films when cobalt sulfate electrolyte is added to the solution, it is established that the dominant phase is the tetragonal phase of $CuBi_2O_4$; however, a significant difference from the observed similar phase for film samples obtained at an applied potential difference of 4.0 V when using electrolyte solution №1 is the broadening of parameter c, which indicates a deformation distortion of the crystal structure (see data in Table 1), which may be associated with the effects of replacing copper with cobalt. At the same time, a change in the synthesis conditions (i.e., a variation in the applied potential difference) in the case of adding cobalt sulfate to the electrolyte solution does not lead to phase change processes, and the structure of the resulting films in the entire studied range of the applied voltage difference is represented by the tetragonal $CuBi_2O_4$ phase. This difference indicates that the addition of cobalt sulfate to the composition of the electrolyte solution leads to the acceleration of bismuth reduction processes at small differences in applied potentials, as well as the release of oxygen, which in turn leads to the formation of a tetragonal $CuBi_2O_4$ phase with varying degrees of structural ordering.

It should also be noted that, when the applied potential difference changes, an increase in the degree of structural ordering is observed, which is expressed not only in a change in the shape of diffraction reflections (the reflections become more symmetrical), but also in a decrease in the parameters of the crystal lattice (see data in Table 1). From this, we can conclude that the addition of cobalt sulfate to the electrolyte solution results in the formation of highly ordered structures, and a change in synthesis conditions is accompanied by an increase in structural ordering.

Figure 4b reveals the assessment results of the change in the elemental composition of the synthesized films with varying applied potential difference, according to which we can conclude that, with an increase in the applied potential difference, copper is partially replaced by cobalt in the composition of the films, the content of which changes from 7.7 at.% at an applied potential difference of 2.0 V to 22.3 at.% at an applied potential difference of 4.0 V. At the same time, no change in the bismuth content in the composition of the films was observed when the synthesis conditions changed, which also indicates that the main substitution is associated with the displacement of copper and its substitution with cobalt. It is also worth noting that, when the applied potential difference changes, a slight decrease in oxygen is observed, the decrease of which may be due to structural ordering.

Figure 2c shows the results of X-ray diffraction of the studied samples of $CuBi/CuBi_2O_4$ films when adding nickel sulfate electrolyte to the solution, obtained by changing the applied potential difference during the synthesis process. The general appearance of the obtained X-ray diffraction patterns indicates that a change in the applied potential difference leads to the formation of films with an almost X-ray amorphous structure (at small potential differences of 2.0–2.5 V), and in the case of high values of applied potential difference (3.5–4.0 V), well-structured films with a tetragonal type of crystal lattice, characteristic of the $CuBi_2O_4$ phase. At the same time, as in the case of films obtained from an electrolyte solution with the addition of cobalt sulfate (solution №2), the addition of nickel sulfate to the electrolyte solution also leads to structural ordering, which is most pronounced in changes in the structural parameters of the resulting films (see data presented in Table 1).

The data on changes in the elemental composition of the films presented in Figure 4c also indicate that the main substitution with a change in synthesis conditions (an increase in the applied potential difference) is associated with the displacement of copper and its substitution with nickel, the content of which also increases from 6.5 at.% at an applied potential difference of 2.0 V to 19.8 at.% at an applied potential difference of 4.0 V, which has a similar trend in the change in the elemental composition for films obtained from solution–electrolyte №2 (with the addition of cobalt sulfate to the electrolyte composition).

Analyzing the obtained changes in the elemental and phase composition of films obtained from solutions–electrolytes №2 and №3, we can conclude that an increase in the applied potential difference leads to the substitution of copper with cobalt or nickel, while new phase inclusions in the composition of the films have not been established.

Figure 5 presents the results of a comparative analysis of the degree of structural ordering (crystallinity degree)—a value that allows one to evaluate the perfection of the crystal structure of the resulting films, as well as the concentration of defective or disordered inclusions, a large number of which can lead to a decrease in the stability of films during their operation. The degree of crystallinity was assessed using a method based on approximating the obtained X-ray diffraction patterns with the required number of pseudo-Voigt functions in order to determine the ratio of the areas of diffraction reflections and the background area. By comparing these values, the degree of crystallinity was determined, i.e., structural ordering of the crystal structure of the obtained samples depending on the conditions of their preparation.

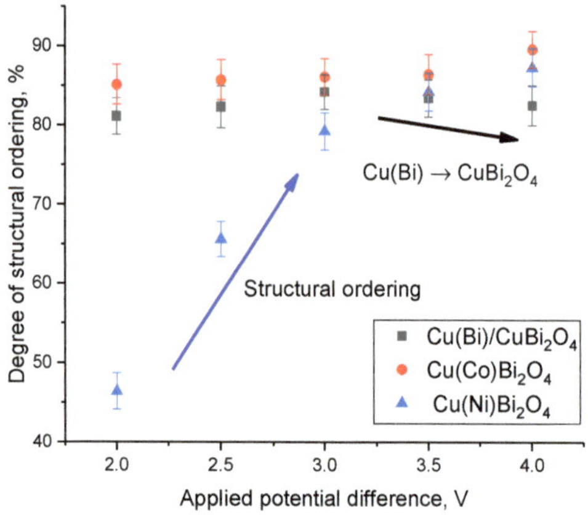

Figure 5. Results of the structural ordering degree of $CuBi/CuBi_2O_4$ films depending on variations in synthesis conditions.

As can be seen from the presented data, the most pronounced changes in the degree of structural ordering are observed for $CuBi/CuBi_2O_4$ films obtained from solution–electrolyte №3, for which, at small differences in applied potentials (2.0–2.5 V), the structural ordering degree has rather low values (less than 70%), which indicates a close to amorphous-like structure of the resulting films. Moreover, in the case of growth in the applied potential difference above 3.0 V, a more than twofold increase in the structural ordering degree for these films is observed in comparison with the results obtained with an applied potential difference of 2.0–2.5 V. The change in the structural ordering degree is close to linear for $CuBi/CuBi_2O_4$ films obtained from solution–electrolyte №3, for which, according to X-ray diffraction data, structural ordering and compaction of the crystal lattice are observed (a decrease in its volume and parameters). In the case of using solution–electrolyte №1, there is a slight decrease in the degree of structural ordering for samples obtained at applied potential differences of 3.5–4.0 V, which is due to phase transformation processes such as Cu(Bi) → $CuBi_2O_4$.

Figure 6 reveals the results of determination of UV–Vis optical transmission spectra of synthesized films depending on the composition of the electrolyte used for synthesis. The general appearance of the obtained spectra is characterized by the presence of a fundamental absorption edge in the region of 300–350 nm, as well as a transmission spectrum in the visible and near-IR ranges. In the case of CuBi/$CuBi_2O_4$ films, the optical spectra are straight with a very low transmission intensity (less than 1%), the value of which indicates the absence of transmission and complete absorption of light in the entire measured range. This nature of the optical spectra indicates the metallic nature of the films obtained, which causes the absence of light transmission. It is important to highlight that the formation of the $CuBi_2O_4$ phase in the films at applied potential differences of 4.0 V leads to a slight increase in the transmission intensity, which may be due to phase changes in the films. In the case of addition of cobalt to the composition of the films, an elevation in transmission intensity is observed in the region above 500 nm, and in the case of the IR range, characteristic interference bands for the PET polymer film, which is the substrate for the synthesized films, are observed. A growth in cobalt content, meanwhile, results in a transmission intensity reduction, which is due to the metallization effect associated with the replacement of copper by cobalt, as well as changes in structural features and degree of crystallinity. For films obtained using an electrolyte with the addition of nickel sulfate, in the case where the films are of an X-ray amorphous nature, the transmittance in the visible and near-IR ranges is rather high, which indicates a direct influence of the degree of structural ordering on the optical properties of the synthesized films. In the case when Cu(Ni)Bi_2O_4 films become structurally ordered (at differences in applied potentials above 3.0 V), the optical transmittance decreases sharply.

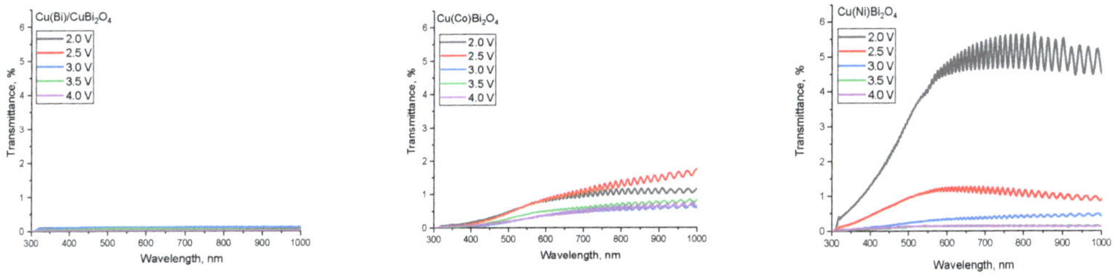

(a) (b) (c)

Figure 6. Measurement results of UV-Vis transmission spectra of synthesized films: (**a**) when using the electrolyte composition to obtain CuBi/$CuBi_2O_4$ films; (**b**) when using the electrolyte composition to obtain Cu(Co)Bi_2O_4 films; (**c**) when using the electrolyte composition to obtain Cu(Ni)Bi_2O_4 films (data are shown on the same scale in order to compare the obtained spectra in terms of transmission intensity).

Figure 7a–c present the study results of the morphological features of the synthesized CuBi/$CuBi_2O_4$ films depending on variations in synthesis conditions, as well as the type of solutions–electrolytes when nickel and cobalt sulfates are added to them. The data are presented in the form of 3D images of the surface of the samples, which reflect changes in both the shape of the grains (their sizes) and the packing density of the grains, expressed in the formation of agglomerates, the presence of which is characteristic of the formation of structures under various synthesis conditions.

The general appearance of the morphological features of CuBi/$CuBi_2O_4$ films obtained from solution–electrolyte №1 indicates that a change in the applied potential differences leads to the enlargement of the grains from which the films are formed, and also that, under all selected synthesis conditions, the surface of the resulting films is rather homogeneous (without large differences in profile heights and fairly low roughness values ~10–20 nm). It follows from this that the use of the selected synthesis conditions is accompanied by the uniform growth of films, without any irregularities, and the deposition process itself is characterized by the formation of spherical or globular particles, the sizes of which vary depending on the synthesis conditions, and, as a consequence, the elemental composition of the resulting structures. The enlargement of grains during the synthesis process can be explained by the effects of accelerated nucleation, characteristic of electrochemical synthesis at large applied potential differences [34].

In the case of samples of the studied CuBi/$CuBi_2O_4$ films obtained by the addition of cobalt sulfate electrolyte to a solution, changing the applied potential difference does not have a significant effect on the morphological features of the grains (no major changes in their sizes have been established); however, at large potential differences (above 3.0 V), the appearance of heterogeneities on the surface is observed, which indicates that the deposition of coatings is uneven (the difference is about 40–70 nm).

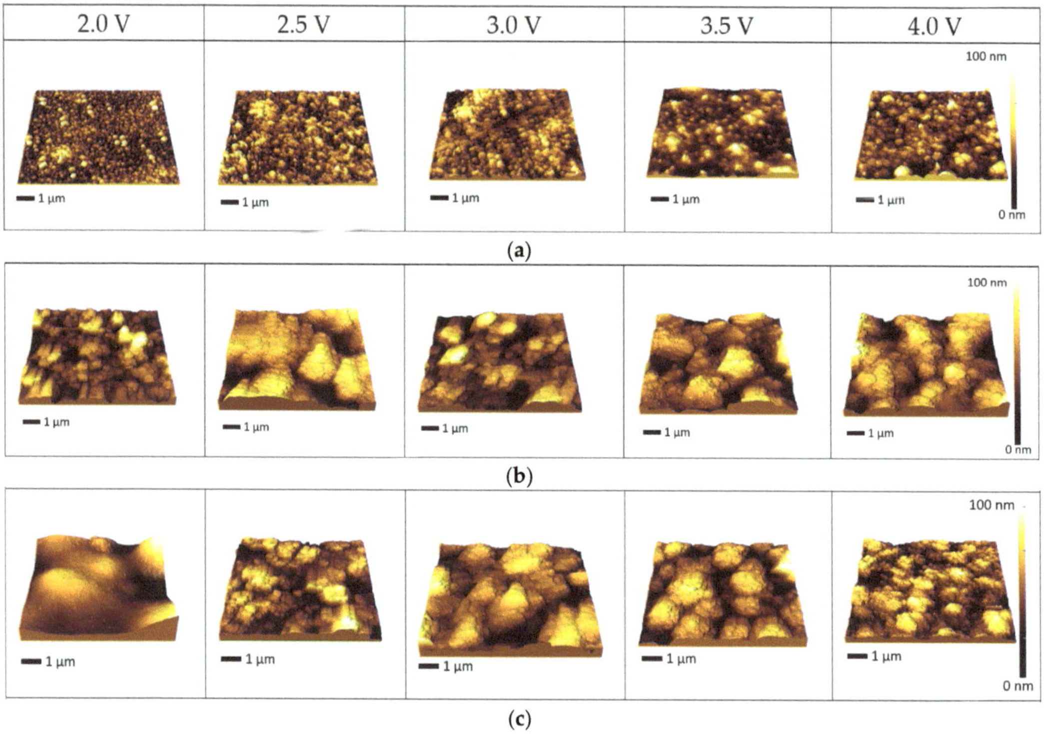

Figure 7. Three-dimensional images of the surface of the studied films: (**a**) when using the electrolyte composition to obtain CuBi/$CuBi_2O_4$ films; (**b**) when using the electrolyte composition to obtain Cu(Co)Bi_2O_4 films; (**c**) when using the electrolyte composition to obtain Cu(Ni)Bi_2O_4 films.

During the electrochemical deposition of CuBi/$CuBi_2O_4$ films obtained by adding a nickel sulfate electrolyte to a solution at an applied potential difference of 2.0 V, the formation of a fine-grained inhomogeneous structure is observed, which has good agreement with the results of X-ray diffraction, characterized by a structure close to X-ray amorphous in the resulting films, which can be explained by the formation of fine grains (the size of which is no more than 5 nm). A rise in the applied potential difference for these films leads to the formation of larger grains; however, as in the case of films obtained from solution-electrolyte №2 (with the addition of cobalt sulfate), the resulting films are characterized by a fairly developed heterogeneous surface, presented in the form of large agglomerates of grains, the average size of which is about 1–2 microns in diameter.

Analyzing the data on changes in the morphological features of the resulting $CuBi/CuBi_2O_4$ films from different compositions of solutions–electrolytes, we can conclude that the formation of the tetragonal $CuBi_2O_4$ phase in the film structure leads to the enlargement of grains, and in the case of addition of cobalt or nickel sulfates to the composition of solutions–electrolytes, it leads to the formation of films with a fairly heterogeneous developed surface, which can be used as anti-friction coatings (resistant to wear).

3.2. Determination of the Influence of Variations in the Composition of Solutions–Electrolytes on the Strength Characteristics of $CuBi/CuBi_2O_4$ Films

Determination of strength characteristics, as well as determination of the effect of addition of cobalt or nickel electrolyte to the solution, which is accompanied by a change in the structural features of the synthesized films, was carried out using the indentation method, the results of which are presented in Figure 8a. These measurements were carried out from different areas of the films under study in order to determine the uniformity of strength characteristics, as well as determine the measurement error and standard deviation.

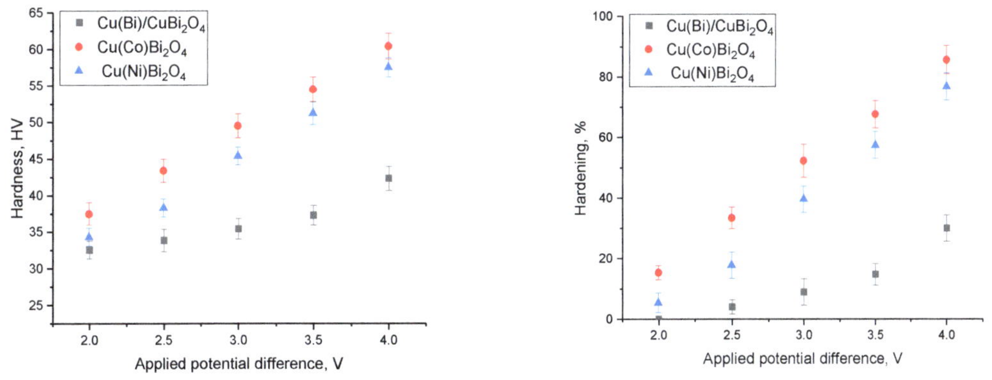

(**a**) (**b**)

Figure 8. (**a**) Results of changes in the hardness of $CuBi/CuBi_2O_4$ films obtained under different production conditions; (**b**) Results of strengthening of synthesized $CuBi/CuBi_2O_4$ films obtained under different production conditions.

The presented dependences of changes in the hardness of $CuBi/CuBi_2O_4$ films obtained under different production conditions can be divided into two types: the first type of changes is associated with the influence of variations in solutions–electrolytes used for film synthesis; the second type with changes in the applied potential difference, which, according to the presented X-ray diffraction data, is accompanied by a change in the degree of structural ordering (when using solutions–electrolytes №2 and №3), as well as phase transformations such as $Cu(Bi) \rightarrow CuBi_2O_4$ (when using solution–electrolyte №1 at potential differences above 3.5 V).

In the case of changes in the hardness values of film samples of the first type, we can conclude that the addition of cobalt or nickel sulfate to the electrolyte solutions leads to an increase in hardness values, which are most pronounced at high potential differences. This increase in hardness can be explained by phase changes in the films, which consist in the fact that, when using solutions–electrolytes №2 and №3, the phase composition of the films is presented in the form of a tetragonal $CuBi_2O_4$ phase, with partial substitution of copper by cobalt or nickel, while the samples obtained from solution–electrolyte №1 at applied potential differences from 2.0 to 3.0 V are represented by the cubic Cu phase, in which some of the atomic positions are occupied by bismuth.

The second type of changes in the hardness values of the film samples under study is associated with the influence of the degree of structural ordering, as well as changes in the elemental composition of the resulting films (an increase in the bismuth content in the case of using electrolyte solution №1 and partial substitution of copper with cobalt or nickel when using electrolyte solutions №2 and №3, respectively). In this case, an increase in these elements in the composition of films leads to a change in hardness (increase in hardness), which is most pronounced at large differences in applied potentials. The change in hardness with a change in the elemental and phase composition of the films is associated with structural changes, which are characterized by the effects of structural ordering (see data in Table 1).

Based on the obtained data on changes in hardness values, hardening factors were calculated, which characterize the effect of changes in strength parameters depending on changes in the degree of structural ordering and phase composition of $CuBi/CuBi_2O_4$ films obtained under different production conditions. The hardening factor was calculated for $CuBi/CuBi_2O_4$ films obtained under different synthesis conditions (in the case of variation in the applied potential difference) by comparative analysis of changes in the hardness values of the samples with the hardness data obtained for samples of $CuBi/CuBi_2O_4$ films synthesized from solution–electrolyte №1 at a potential difference of 2.0 V. The results of the strength characteristics assessment are presented in Figure 8b.

The general appearance of the presented changes in the results of strengthening indicates that the greatest influence on the increase in strength characteristics is exerted by the effect of partial substitution of copper with cobalt or nickel, as well as an increase in their concentration in the films, which leads to more than 1.5-fold strengthening of the films. In the case of films obtained from a solution–electrolyte №1, the change in hardness (i.e., hardening) is most pronounced during phase transformations of the $Cu(Bi) \rightarrow CuBi_2O_4$ type, which lead to film strengthening by more than 20–25%, in comparison with films obtained at lower potential differences (below 3.0 V).

Figure 9 shows the results of a comparative analysis of the factors of structural ordering and film strengthening (hardness changes) depending on the type of films obtained. As can be seen from the presented dependence, the most pronounced effect of structural ordering on strengthening is manifested for $CuBi/CuBi_2O_4$ films obtained from solution–electrolyte №3 (with the addition of nickel sulfate), the use of which, at potential differences equal to 2.0 and 2.5 V, leads to the formation of films with an almost X-ray amorphous structure, the ordering of which leads to a sharp change in strength characteristics. In the case of $CuBi/CuBi_2O_4$ films obtained from solution–electrolyte №1, the main contribution to strengthening is made by phase transformation processes, which manifest themselves at applied potential differences above 3.0 V (in this case, the structural ordering degree for these films decreases due to phase transformations accompanied by deformation distortion of the crystal lattice due to rearrangement of the crystal structure and the formation of the tetragonal $CuBi_2O_4$ phase).

One of the key factors characterizing the wear resistance of materials is their resistance to long-term mechanical stress due to friction or pressure. To assess wear resistance, as a rule, tribological methods are used, which make it possible to determine such quantities as the coefficient of dry friction, wear rate or mass loss of samples during long-term life tests.

Figure 10 shows the results of tribological tests of the studied $CuBi/CuBi_2O_4$ films carried out with sequential exposure of the indenter to the surface of film samples. Based on the tribological test data, the dry friction coefficient was calculated, the value of which was measured after each 1000 consecutive tests. The general appearance of the presented dependences of changes in the dry friction coefficient when changing the type of solution–electrolyte for producing $CuBi/CuBi_2O_4$ films can be characterized as follows. The use of solutions–electrolytes №2 and №3 leads to a slight increase in the coefficient of dry friction of films in the initial state from 0.24 to 0.25 in comparison with the value of the coefficient of dry friction for $CuBi/CuBi_2O_4$ films obtained from solution–electrolyte №1 equal to 0.17. This increase can be explained by effects associated with changes in the morphological features of the resulting films when using solutions–electrolytes №2 and №3, for which, according to the results of atomic force microscopy presented in Figure 4a–c, an increase in the surface roughness of the films and an enlargement of their sizes are observed. Moreover, for all three types of films, regardless of the applied potential differences, no significant differences in the initial value of the dry friction coefficient were observed.

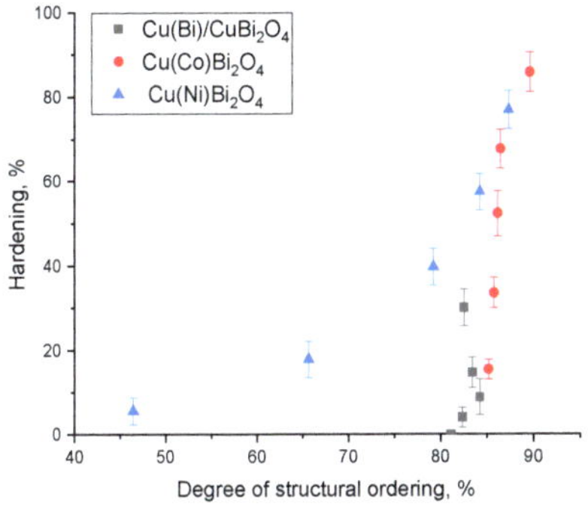

Figure 9. Results of a comparative analysis of factors of structural ordering and strengthening.

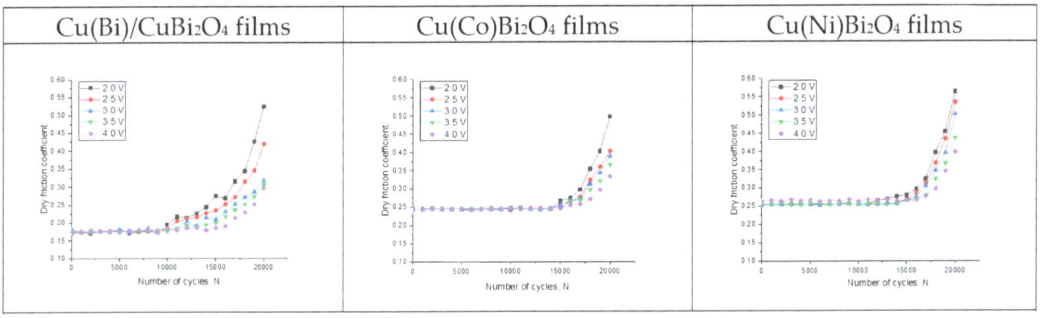

(a)

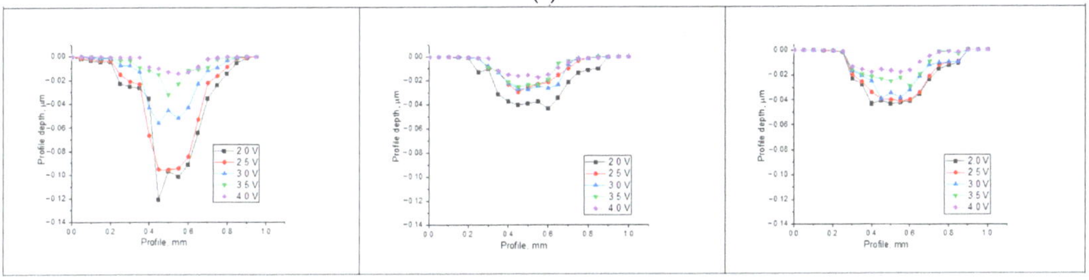

(b)

Figure 10. (**a**) Results of measurement of dry friction coefficient depending on the number of test cycles; (**b**) Results of evaluation of wear profile of film samples depending on their production conditions.

During long-term tests, a change in the dry friction coefficient (its increase) indicates surface degradation and wear, which leads to the creation of additional obstacles for the moving indenter and increases friction.

As a rule, wear-resistant coatings are able to withstand a fairly large number of test cycles (about 10,000–15,000), which characterizes their resistance to external influences. In the case of $CuBi/CuBi_2O_4$ film samples obtained using electrolyte solution №1, the main changes in the dry friction coefficient are observed after 10,000 cycles, while the most pronounced changes occur after 15,000 cycles and consist of a sharp deterioration in the coefficient (see data in Figure 11a). At the same time, the most pronounced changes are observed for films obtained at potential differences of 2.0–3.0 V, in which the cubic phase of copper dominates, which has a fairly low resistance to degradation during wear resistance tests. In turn, phase transformations of the type Cu(Bi) $\rightarrow$ $CuBi_2O_4$, which occur at applied potential differences above 3.5 V (in the case of using electrolyte solution №1), lead to a decrease in the degradation of the dry friction coefficient in comparison with the initial values, which indicates an increase in the resistance of films to long-term mechanical influences (friction) in the case wherein the phase composition of the films is represented by the tetragonal $CuBi_2O_4$ phase, which has higher hardness values than copper coatings with partial substitution of copper by bismuth.

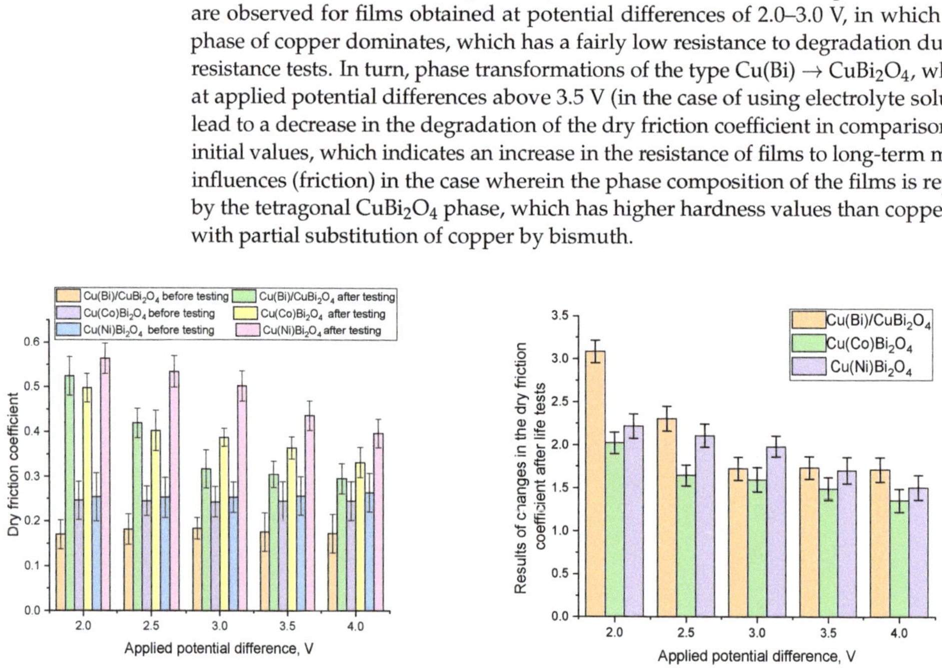

(a) (b)

Figure 11. (**a**) Results of a comparative analysis of changes in the value of the dry friction coefficient before and after tribological tests; (**b**) Results of evaluation of changes in the dry friction coefficient after the tribological life tests compared to the initial value (this change reflects the dry friction coefficient degradation degree during the tests).

In the case of films obtained using solutions–electrolytes №2 and №3, changes in the dry friction coefficient occur after 15,000 consecutive tests, and the degradation of the coefficient is about 1.5–2.0 times, while the degradation of the dry friction coefficient for films obtained using solution–electrolyte №1 is more than 2.5–3.0 times in comparison with the initial data (see data in Figure 11a,b).

Thus, analyzing the obtained results of changes in the value of the dry friction coefficient, we can conclude that the formation of films with a phase composition presented in the form of a tetragonal $CuBi_2O_4$ phase leads to an increase in wear resistance and maintaining resistance to degradation during friction over a long number of cyclic tests, and in the case of partial substitution of copper with cobalt or nickel, an increase in wear resistance and a decrease in degradation of the film surface during tribological tests are observed.

Figure 10b shows the results of assessing the wear profile of the studied $CuBi/CuBi_2O_4$ films depending on the production conditions, as well as the electrolyte solutions used (with the addition of cobalt or nickel sulfates). These profiles were obtained after conducting tribological tests for resistance to friction, and the profiles themselves reflect changes in the resistance of films to mechanical degradation caused by external mechanical influences (in this case, friction).

According to the data obtained, it is clear that the greatest changes associated with the degradation of the film surface are observed for copper films (samples obtained using solution–electrolyte №1 at potential differences from 2.0 to 3.0 V). At the same time, the assessment of wear profiles (i.e., depth and width of the profile) for samples obtained using solution–electrolyte №1 indicates that an increase in the concentration of bismuth in the composition of the films leads to an increase in wear resistance, which also manifests itself for samples obtained from solutions–electrolytes №2 and №3 when their elemental composition changes (when copper is replaced by cobalt or nickel). Moreover, in the case of films obtained from solutions–electrolytes №2 and №3, the wear profiles have a significantly shallower depth, in comparison with similar profile data obtained for films synthesized using solution–electrolyte №1, which indicates their high resistance to external influences, in particular to the loss of sample mass during prolonged friction.

Based on the obtained data on changes in the dry friction coefficient, as well as wear profiles, the wear rate was calculated during tribological tests of the surface of films obtained from various solutions–electrolytes and when changing synthesis conditions (variations in the applied potential difference). The evaluation results are presented in Figure 12 in the form of a comparison chart.

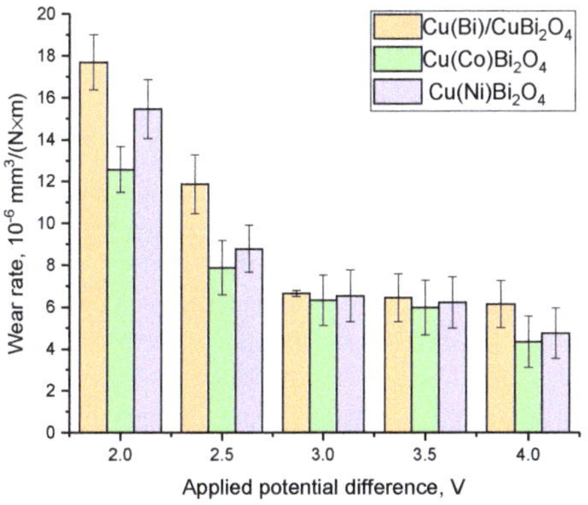

Figure 12. Results of film wear rate assessment during tribological tests.

According to the data obtained, the films obtained at an applied potential difference of 2.0 for all three types of solutions–electrolytes have the highest degradation rate (surface wear rate during life-long tribological tests), which is due to structural (in the case of films obtained from a solution–electrolyte №1, a high concentration of copper) or morphological features (in the case of films obtained from solution–electrolyte №3, an amorphous-like structure consisting of small grains), which leads to the accelerated degradation of films. Moreover, in the case when the concentration of cobalt or nickel in the film composition increases (when using solutions–electrolytes №2 and №3), as well as bismuth (when using a solution–electrolyte №1), a more than 2.0–2.5-fold decrease in the surface wear rate is observed, which is expressed in lower changes in the dry friction coefficient, as well as smaller wear profiles.

Thus, summing up the results of the measurements of the strength and tribological characteristics of the films under study, we can conclude that the substitution of copper with cobalt or nickel that occurs in the structure of the films leads to the possibility of obtaining high-strength wear-resistant films that are highly resistant to long-term mechanical influences.

Resistance to mechanical influences, in particular mechanical pressure, shock and friction, is one of the important parameters when using protective materials in real conditions. Thus, good wear resistance indicators allow for the use of these protective coatings during mechanical friction in the case when moving parts are used that can come into contact with each other during operation. In this case, high resistance to wear due to friction makes it possible to eliminate the factors of destruction of coatings during prolonged mechanical action, as well as to avoid the effects associated with a decrease in shielding efficiency in the case of partial or complete separation of coatings from the surface. Comparing the hardness values of the synthesized coatings with similar nitride (TiN) coatings obtained by magnetron sputtering [35,36] or by vapor deposition [37], they are approximately 1.5–2.0 times lower than the hardness of TiN coatings [36], and 7–10 times lower than coatings obtained by vapor deposition [37]. In comparison with carbide coatings [38,39], the hardness values of CuBi/$CuBi_2O_4$, Cu(Co)Bi_2O_4 and Cu(Ni)Bi_2O_4 films are also approximately 5–6 times lower. However, comparing wear resistance indicators determined from changes in the dry friction coefficient, we can conclude that the synthesized films under consideration have comparable wear resistance indicators to carbide and nitride coatings during long-term wear tests. This difference is due to the peculiarities of the coating production methods, as well as differences in the scope of application. As a rule, nitride coatings are used to protect against corrosion and degradation of steel structures, but do not have high shielding characteristics.

3.3. Determination of Gamma Ray Shielding Efficiency Using Different Films

To determine the shielding efficiency of gamma radiation with energies of 0.13, 0.66 and 1.23 MeV generated by Co^{57}, Cs^{137}, Na^{22} sources, respectively, five samples of CuBi/$CuBi_2O_4$ films obtained under different conditions were selected: Cu films with a low bismuth content, obtained from electrolyte solution №1 at an applied potential difference of 2.0 V, Cu(Bi) films with a bismuth content of about 20 at.%, obtained from an electrolyte solution at an applied potential difference of 3.5 V, $CuBi_2O_4$ films obtained from electrolyte solution №1 at a potential difference of 4.0 V, Cu(Co)Bi_2O_4 and Cu(Ni)Bi_2O_4 films obtained at an applied potential difference of 4.0 V from electrolyte solutions №2 and №3, respectively. Shielding experiments were carried out according to a standard scheme; the efficiency was assessed using the calculation formulas used in [40–42], which make it possible to obtain the values of the shielding characteristics of films, which were compared with the calculated values performed using the XCOM code, the simulation results of which are presented in Figure 13.

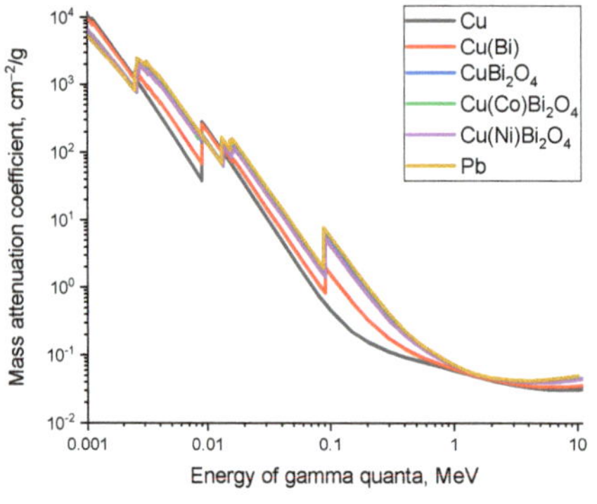

Figure 13. Calculation results of the mass attenuation coefficient obtained using the XCOM program code (data on changes in the mass attenuation coefficient for lead are also given as an example).

The general appearance of the presented dependences of the change in the mass attenuation coefficient indicates the presence of local absorption maxima and minima, characteristic of the structural features of the films, as well as their electronic structure. At the same time, the analysis of the obtained dependences showed high values of the absorbing (shielding) ability of the selected objects of study in the region of low energies of gamma quanta (less than 0.1 MeV), characteristic of the processes of interaction of gamma quanta through the mechanisms of the photoelectric effect. At the same time, no significant differences are observed in the energy region of more than 1 MeV, since in this region the dominant role is played by the processes of formation of electron–positron pairs, the formation of which is accompanied by the effects of the formation of secondary radiation. Also shown in Figure 10 is the dependence of the mass attenuation coefficient for lead, which has the maximum shielding efficiency for gamma radiation among all currently known protective shielding materials. As can be seen from a comparative analysis of the dependences of the mass attenuation coefficient for the films under study with data for lead, the most pronounced changes are observed for samples in the energy range of gamma radiation up to 1 MeV, where the processes of the photoelectric effect and the Compton effect dominate, while in the region of high-energy gamma quanta (with energy of more than 1.0 MeV), the dependences have almost the same range of values.

The assessment results of the shielding characteristics (mass and linear attenuation coefficient, half-attenuation thickness and mean free path) are presented in Table 2. The general appearance of the presented dependences of the shielding characteristics indicates the positive influence of phase transformations such as Cu(Bi) → $CuBi_2O_4$, as well as the substitution of copper with cobalt and nickel in the composition of $Cu(Co)Bi_2O_4$ and $Cu(Ni)Bi_2O_4$ films, which leads to an increase in the efficiency of shielding characteristics by more than 3.0–4.0 times when shielding gamma quanta with energies of 0.13 MeV in comparison with Cu films, which have minimal gamma-ray shielding efficiency indicators. In the case of shielding gamma rays with energies of 0.66 and 1.23 MeV, the difference in the efficiency of shielding characteristics in comparison with Cu films is about 1.5–2.0 times, which is due to differences in the mechanisms of interaction of gamma quanta with higher energies, which are accompanied by the formation of secondary radiation.

Table 2. Data on the shielding characteristics of the thin film samples under study.

Sample Type	Shielding Parameter for Gamma Rays with an Energy of 0.13 MeV (Co^{57})			
	MAC, cm^2/g	LAC, cm^{-1}	$\Delta_{1/2}$, cm^{-1}	MFP, cm
Cu *	0.39	3.47	0.20	0.29
Cu(Bi)	1.42	11.84	0.06	0.08
$CuBi_2O_4$	1.59	13.63	0.05	0.07
$Cu(Co)Bi_2O_4$	1.85	16.01	0.04	0.06
$Cu(Ni)Bi_2O_4$	1.73	14.93	0.05	0.07
Sample Type	**Shielding Parameter for Gamma Rays with an Energy of 0.66 MeV (Cs^{137})**			
	MAC, cm^2/g	LAC, cm^{-1}	$\Delta_{1/2}$, cm^{-1}	MFP, cm
Cu	0.067	0.60	1.16	1.68
Cu(Bi)	0.075	0.63	1.11	1.61
$CuBi_2O_4$	0.105	0.90	0.77	1.11
$Cu(Co)Bi_2O_4$	0.113	0.98	0.71	1.02
$Cu(Ni)Bi_2O_4$	0.111	0.96	0.72	1.04
Sample Type	**Shielding Parameter for Gamma Rays with an Energy of 1.23 MeV (Na^{22})**			
	MAC, cm^2/g	LAC, cm^{-1}	$\Delta_{1/2}$, cm^{-1}	MFP, cm
Cu	0.039	0.35	2.01	2.88
Cu(Bi)	0.048	0.40	1.73	2.51
$CuBi_2O_4$	0.052	0.45	1.56	2.24
$Cu(Co)Bi_2O_4$	0.058	0.50	1.38	1.99
$Cu(Ni)Bi_2O_4$	0.054	0.47	1.49	2.15

* Values are given to reflect the low efficiency of using Cu films as protective shielding materials.

The data on alterations in shielding characteristics presented in Table 2 demonstrate that the addition of cobalt to films leads to an increase in MAC efficiency by 15–16% in comparison with $CuBi_2O_4$ films and about 30% in comparison with Cu(Bi) films. In the case of adding nickel to the film composition, the efficiency of MAS increases by 8–10% when compared with $CuBi_2O_4$ films and by about 21–22% when compared with Cu(Bi) films Thus, it can be concluded that the addition of cobalt and nickel to the composition, which makes it possible to obtain films with the structure of $Cu(Co)Bi_2O_4$ and $Cu(Ni)Bi_2O_4$, results in an elevation in the shielding efficiency of the order of 10–20%. Moreover, these films have higher levels of mechanical strength and wear resistance to external influences, which together allows them to be used as protective coatings exposed to external mechanical influences, friction during use, etc.

Figure 14a presents the results of a comparative analysis of the experimentally determined values of the mass attenuation coefficient of the films selected as objects of study with the simulation results performed using the XCOM code. According to the data presented, there is good agreement between the experimental and calculated values of the mass attenuation coefficient (the difference is no more than 10%), which indicates that the obtained experimental values of the shielding characteristics reliably reflect the effectiveness of the synthesized films.

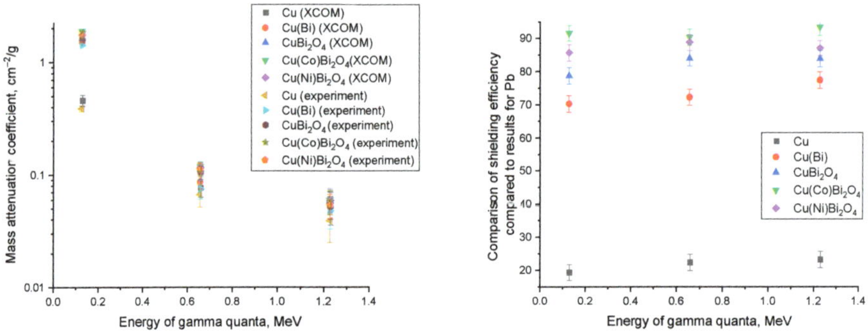

(a) (b)

Figure 14. (**a**) Results of a comparative analysis of the mass attenuation coefficient obtained experimentally and using the modeling method in the XCOM code; (**b**) Results of evaluation of the shielding efficiency of $CuBi/CuBi_2O_4$ films in comparison with the mass attenuation coefficient of Pb films obtained on the basis of calculated data using the XCOM program code.

Figure 14b presents the results of the shielding efficiency (mass attenuation coefficient) of the films under study for all three types of gamma ray energy in comparison with the value of the mass attenuation coefficient of lead obtained from the simulation results using the XCOM code. As can be seen from the presented data, the synthesized $CuBi_2O_4$ films obtained from solution–electrolyte №1 at a potential difference of 4.0 V, $Cu(Co)Bi_2O_4$ and $Cu(Ni)Bi_2O_4$ films, have a shielding efficiency of the order of 0.8–0.9 of the value of the shielding characteristics of lead; however, the density of the films is significantly lesser than the density of lead, and as a result, the use of these films as shielding protective materials will reduce the weight and overall dimensions by reducing the weight of the shields, without losing shielding efficiency (the reduction in shielding efficiency when using these films will be no more than 10–20%). It is also worth to note that the use of the effect of replacing copper with cobalt or nickel in the composition of the tetragonal phase of $CuBi_2O_4$ leads to an increase in the shielding efficiency by 5–10% in comparison with samples of $CuBi_2O_4$ films without substitution, which also indicates the positive effect of using dopants to modify the resulting films.

4. Conclusions

In the course of the studies carried out using the X-ray diffraction method, results were obtained on changes in the phase composition of films with variations in synthesis conditions (differences in applied potentials) and changes in the compositions of solutions–electrolytes (with the addition of cobalt and nickel sulfates). In the case of using solution–electrolyte №1 (without any additives), a change in the applied potential difference leads to the initialization of phase transformation processes such as Cu(Bi) → $CuBi_2O_4$, which occur when the applied potential differences are above 3.0 V, as well as when the bismuth content in the films is about 20 at.% and higher. In this case, in the case of small applied potential differences (below 3.0 V), the dominance of the cubic copper phase was observed in the structure of the films, the change in the structural parameters for which indicates the processes of partial substitution of copper with bismuth (substitution or interstitial phase).

When using solutions–electrolytes №2 and №3 (addition of cobalt sulfate and nickel sulfate, respectively), the dominant phase in the film structure is the tetragonal $CuBi_2O_4$ phase, for which a change in the applied potential difference during film deposition leads to an increase in structural ordering, as well as a decrease in the crystal lattice parameters, indicating film densification. At the same time, analysis of these changes in the elemental composition of films when using solutions–electrolytes №2 and №3 indicates that the main changes are associated with the substitution of copper with cobalt or nickel (depending on the type of electrolyte solution used), the concentrations of which grow with an increase in the applied potential differences.

The results of studies of the strength and tribological characteristics of the synthesized films, depending on the conditions for their production, showed the following. It has been experimentally established that the addition of cobalt or nickel sulfates to the composition of solutions–electrolytes №2 and №3 leads to an increase in the strength of the resulting films from 20 to 80%, depending on the production conditions (with variations in the difference in applied potentials). In the course of determining the tribological characteristics, it was found that, when replacing copper with cobalt or nickel, depending on the type of solution–electrolyte used, as well as synthesis conditions (in the case of varying the difference in applied potentials), an increase in resistance to wear is observed, as well as a decrease in the wear profile, which indicates an increase in wear resistance and degradation to external mechanical influences. At the same time, alterations in the morphological features of $CuBi/CuBi_2O_4$ films depending on variations in synthesis conditions (changes in the electrolyte solution or the difference in applied potentials) do not lead to significant changes in the dry friction coefficient.

When determining the shielding efficiency of gamma radiation, it was found that replacing copper with cobalt or nickel in the composition of $CuBi_2O_4$ films leads to an increase in the shielding efficiency of low-energy gamma radiation by 3.0–4.0 times in comparison with copper films, and 1.5–2.0 times for high-energy gamma quanta. The efficiency decrease in this case is due to differences in the mechanisms of interaction of gamma quanta, as well as the occurrence of secondary radiation as a result of the formation of electron–positron pairs and the Compton effect. It was found that the use of the effect of substitution of copper with cobalt or nickel in the composition of the tetragonal phase of $CuBi_2O_4$ results in an elevation in the shielding efficiency by 5–10% in comparison with samples of $CuBi_2O_4$ films without substitution, which also indicates the positive effect of using dopants to modify the resulting films.

Author Contributions: Conceptualization, M.T.I., A.L.K., D.I.S. and D.B.K.; methodology, M.T.I., A.L.K., D.I.S. and D.B.K.; formal analysis, M.T.I., A.L.K., D.I.S. and D.B.K.; investigation, M.T.I., A.L.K., D.I.S. and D.B.K.; resources, M.T.I., A.L.K., D.I.S. and D.B.K.; writing—original draft preparation, review, and editing, M.T.I., A.L.K., D.I.S. and D.B.K.; visualization, M.T.I., A.L.K., D.I.S. and D.B.K.; supervision, D.B.K. All authors have read and agreed to the published version of the manuscript.

Funding: This study was funded by the Science Committee of the Ministry of Science and Higher Education of the Republic of Kazakhstan (No. AP14871152).

Data Availability Statement: The original contributions presented in the study are included in the article, further inquiries can be directed to the corresponding author.

Conflicts of Interest: The authors declare no conflicts of interest.

References

1. National Research Council; Division on Earth and Life Studies; Nuclear and Radiation Studies Board; Committee on Radiation Source Use and Replacement. *Radiation Source Use and Replacement: Abbreviated Version*; National Academies Press: Washington, DC, USA, 2008; pp. 1–100.
2. Kottou, S.; Nikolopoulos, D.; Vogiannis, E.; Koulougliotis, D.; Petraki, E.; Yannakopoulos, P.H. How safe is the environmental electromagnetic radiation. *J. Phys. Chem. Biophys.* **2014**, *4*, 1000146. [CrossRef]
3. Panagopoulos, D.J.; Margaritis, L.H. 2. Theoretical Considerations for the Biological Effects of Electromagnetic Fields. In *Biological Effects of Electromagnetic Fields: Mechanisms, Modeling, Biological Effects, Therapeutic Effects, International Standards, Exposure Criteria*; Springer: Berlin/Heidelberg, Germany, 2013; p. 2.
4. Classic, K.; Le Guen, B.; Kase, K.; Vetter, R. Safety and radiation protection culture. In *Radiological Safety and Quality: Paradigms in Leadership and Innovation*; Springer: Dordrecht, The Netherlands, 2014; pp. 263–277.
5. Gutierrez, J.M.; Emery, R.J. A 30-year radiation safety prospectus describing organizational drivers, program activities, and outcomes. *Health Phys.* **2022**, *122*, 352–359. [CrossRef] [PubMed]
6. Moore, Q.T. Validity and reliability of a radiation safety culture survey instrument for radiologic technologists. *Radiol. Technol.* **2021**, *92*, 547–560. [PubMed]
7. Morris, V. A quality educational program can significantly improve radiation safety. *Health Phys.* **2003**, *84*, S71–S73. [CrossRef] [PubMed]
8. Mirzayev, M.N.; Popov, E.; Demir, E.; Abdurakhimov, B.A.; Mirzayeva, D.M.; Sukratov, V.A.; Mutali, A.K.; Tiep, V.N.; Biira, S.; Tashmetov, M.Y.; et al. Thermophysical behavior of boron nitride and boron trioxide ceramics compounds with high energy electron fluence and swift heavy ion irradiated. *J. Alloys Compd.* **2020**, *834*, 155119. [CrossRef]
9. Abdelbagi, H.A.; Skuratov, V.A.; Motloung, S.V.; Njoroge, E.G.; Mlambo, M.; Hlatshwayo, T.T.; Malherbe, J.B. Effect of swift heavy ions irradiation on the migration behavior of strontium implanted into polycrystalline SiC. *Nucl. Instrum. Methods Phys. Res. Sect. B Beam Interact. Mater. At.* **2019**, *451*, 113–121. [CrossRef]
10. Akel, F.Z.; Izerrouken, M.; Belgaid, M. Neutrons and swift heavy ions irradiation induced damage in SiC single crystal. *Mater. Today Commun.* **2023**, *37*, 107268. [CrossRef]
11. Kozlovskiy, A.; Kenzhina, I.; Alyamova, Z.A.; Zdorovets, M. Optical and structural properties of AlN ceramics irradiated with heavy ions. *Opt. Mater.* **2019**, *91*, 130–137. [CrossRef]
12. Zinkle, S.J.; Jones, J.W.; Skuratov, V.A. Microstructure of swift heavy ion irradiated SiC, Si3N4 and AlN. *MRS Online Proc. Libr. (OPL)* **2000**, *650*, R3–R19. [CrossRef]
13. Glyva, V.; Natalia, K.; Nazarenko, V.; Burdeina, N.; Karaieva, N.; Levchenko, L.; Panova, O.; Tykhenko, O.; Khalmuradov, B.; Khodakovskyy, O. Development and study of protective properties of the composite materials for shielding the electromagnetic fields of a wide frequency range. *East.-Eur. J. Enterp. Technol.* **2020**, *2*, 104. [CrossRef]
14. Pikhay, E.; Roizin, Y.; Nemirovsky, Y. Ultra-low power consuming direct radiation sensors based on floating gate structures. *J. Low Power Electron. Appl.* **2017**, *7*, 20. [CrossRef]
15. Schrimpf, R.D.; Fleetwood, D.M. (Eds.) *Radiation Effects and Soft Errors in Integrated Circuits and Electronic Devices*; World Scientific: Singapore, 2004; Volume 34.
16. Malothu, R.; Rao, R.S. Combination of Copper Bismuth Oxide ($CuBi_2O_4$) and Polymer Composites from Plastic Waste: A Boon for EMF Shielding. *I-Manag. J. Future Eng. Technol.* **2021**, *16*, 11.
17. Awad, E.H.; Raslan, H.A.; Abou-Laila, M.T.; Taha, E.O.; Atta, M.M. Synthesis and characterization of waste polyethylene/Bi_2O_3 composites reinforced with CuO/ZnO nanoparticles as sustainable radiation shielding materials. *Polym. Eng. Sci.* 2024; *early view*.
18. Eren Belgin, E.; Aycik, G.A. Preparation and radiation attenuation performances of metal oxide filled polyethylene based composites for ionizing electromagnetic radiation shielding applications. *J. Radioanal. Nucl. Chem.* **2015**, *306*, 107–117. [CrossRef]
19. Cho, J.H.; Kim, M.S.; Rhim, J.D. Comparison of radiation shielding ratios of nano-sized bismuth trioxide and molybdenum. *Radiat. Eff. Defects Solids* **2015**, *170*, 651–658. [CrossRef]
20. Oğul, H.; Agar, O.; Bulut, F.; Kaçal, M.R.; Dilsiz, K.; Polat, H.; Akman, F. A comparative neutron and gamma-ray radiation shielding investigation of molybdenum and boron filled polymer composites. *Appl. Radiat. Isot.* **2023**, *194*, 110731. [CrossRef] [PubMed]
21. Abdolahzadeh, T.; Morshedian, J.; Ahmadi, S. Preparation and characterization of nano WO_3/Bi_2O_3/GO and $BaSO_4$/GO dispersed HDPE composites for X-ray shielding application. *Polyolefins J.* **2022**, *9*, 73–83.
22. Seifert, N.; Slankard, P.; Kirsch, M.; Narasimham, B.; Zia, V.; Brookreson, C.; Vo, A.; Mitra, S.; Gill, B.; Maiz, J. Radiation-induced soft error rates of advanced CMOS bulk devices. In Proceedings of the 2006 IEEE International Reliability Physics Symposium Proceedings, San Jose, CA, USA, 26–30 March 2006.

23. Seifert, N.; Ambrose, V.; Gill, B.; Shi, Q.; Allmon, R.; Recchia, C.; Mukherjee, S.; Nassif, N.; Krause, J.; Pickholtz, J.; et al. On the radiation-induced soft error performance of hardened sequential elements in advanced bulk CMOS technologies. In Proceedings of the 2010 IEEE International Reliability Physics Symposium, Anaheim, CA, USA, 2–6 May 2010.
24. Kim, B.R.; Lee, H.K.; Kim, E.; Lee, S.H. Intrinsic electromagnetic radiation shielding/absorbing characteristics of polyaniline-coated transparent thin films. *Synth. Met.* **2010**, *160*, 1838–1842. [CrossRef]
25. Al-Balushi, M.A.; Ahmed, N.M.; Zyoud, S.H.; Mohammed Ali, M.K.; Akhdar, H.; Aldaghri, O.A.; Ibnaouf, K.H. Ionization radiation shielding effectiveness of lead acetate, lead nitrate, and bismuth nitrate-doped zinc oxide nanorods thin films: A comparative evaluation. *Materials* **2021**, *15*, 3. [CrossRef] [PubMed]
26. Soğuksu, A.K.; Kerli, S.; Kavun, Y.; Alver, Ü. Synthesis and characterizations of Ce-doped ZnO thin films for radiation shielding. *Opt. Mater.* **2024**, *148*, 114941. [CrossRef]
27. Kadyrzhanov, D.B.; Kaliyekperov, M.E.; Idinov, M.T.; Kozlovskiy, A.L. Study of the Structural, Morphological, Strength and Shielding Properties of $CuBi_2O_4$ Films Obtained by Electrochemical Synthesis. *Materials* **2023**, *16*, 7241. [CrossRef] [PubMed]
28. Muthamma, M.V.; Bubbly, S.G.; Gudennavar, S.B.; Narendranath, K.S. Poly (vinyl alcohol)–bismuth oxide composites for X-ray and γ-ray shielding applications. *J. Appl. Polym. Sci.* **2019**, *136*, 47949. [CrossRef]
29. Eskalen, H.; Kavun, Y.; Kerli, S.; Eken, S. An investigation of radiation shielding properties of boron doped ZnO thin films. *Opt. Mater.* **2020**, *105*, 109871. [CrossRef]
30. AbuAlRoos, N.J.; Amin, N.A.B.; Zainon, R. Conventional and new lead-free radiation shielding materials for radiation protection in nuclear medicine: A review. *Radiat. Phys. Chem.* **2019**, *165*, 108439. [CrossRef]
31. Chan, T.-C.; Chueh, Y.-L.; Liao, C.-N. Manipulating the crystallographic texture of nanotwinned Cu films by electrodeposition. *Cryst. Growth Des.* **2011**, *11*, 4970–4974. [CrossRef]
32. Baskaran, I.; Sankara Narayanan, T.S.N.; Stephen, A. Pulsed electrodeposition of nanocrystalline Cu–Ni alloy films and evaluation of their characteristic properties. *Mater. Lett.* **2016**, *60*, 1990–1995. [CrossRef]
33. Otte, H.M. Lattice parameter determinations with an X-ray spectrogoniometer by the debye-scherrer method and the effect of specimen condition. *J. Appl. Phys.* **1961**, *32*, 1536–1546. [CrossRef]
34. Ebrahimi, F.; Liscano, A.J.; Kong, D.; Krishnamoorthy, V. Evolution of texture in electrodeposited Ni/Cu layered nanostructures. *Philos. Mag.* **2003**, *83*, 457–476. [CrossRef]
35. Wu, K.-J.; Edmund, C.M.; Shang, C.; Guo, Z. Nucleation and growth in solution synthesis of nanostructures–from fundamentals to advanced applications. *Prog. Mater. Sci.* **2022**, *123*, 100821. [CrossRef]
36. Chauhan, K.V.; Rawal, S.K. A review paper on tribological and mechanical properties of ternary nitride based coatings. *Procedia Technol.* **2014**, *14*, 430–437. [CrossRef]
37. Liu, Z.; Ren, S.; Li, T.; Chen, P.; Hu, L.; Wu, W.; Li, S.; Liu, H.; Li, R.; Zhang, Y. A Comparison Study on the Microstructure, Mechanical Features, and Tribological Characteristics of TiN Coatings on Ti_6Al_4V Using Different Deposition Techniques. *Coatings* **2024**, *14*, 156. [CrossRef]
38. Wu, S.; Wu, S.; Zhang, G.; Zhang, W. Hardness and elastic modulus of titanium nitride coatings prepared by pirac method. *Surf. Rev. Lett.* **2018**, *25*, 1850040. [CrossRef]
39. Pujada, B.R.; Tichelaar, F.D.; Janssen, G.C.A.M. Hardness of and stress in tungsten carbide–diamond like carbon multilayer coatings. *Surf. Coat. Technol.* **2008**, *203*, 562–565. [CrossRef]
40. Bouzakis, K.-D.; Hadjiyiannis, S.; Skordaris, G.; Mirisidis, I.; Michailidis, N.; Efstathiou, K.; Pavlidou, E.; Erkens, G.; Cremer, R.; Rambadt, S.; et al. The effect of coating thickness, mechanical strength and hardness properties on the milling performance of PVD coated cemented carbides inserts. *Surf. Coat. Technol.* **2004**, *177*, 657–664. [CrossRef]
41. Ezzeldin, M.; Al-Harbi, L.M.; Sadeq, M.S.; Mahmoud, A.E.; Muhammad, M.A.; Ahmed, H.A. Impact of CdO on optical, structural, elastic, and radiation shielding parameters of CdO–PbO–ZnO–B2O3–SiO2 glasses. *Ceram. Int.* **2023**, *49*, 19160–19173. [CrossRef]
42. Sayyed, M.I.; Rammah, Y.S.; Abouhaswa, A.S.; Tekin, H.O.; Elbashir, B.O. ZnO-B_2O_3-PbO glasses: Synthesis and radiation shielding characterization. *Phys. B Condens. Matter* **2018**, *548*, 20–26. [CrossRef]

crystals

MDPI

Article

Design and Study of Composite Film Preparation Platform

Chao Li [1,2], Wenxin Li [1,2,*], Guangqin Wu [1], Guojin Chen [1,2], Junyi Wu [3], Niushan Zhang [4], Yusen Gan [1,2], Dongqi Zhang [1,2] and Chang Chen [1,2]

[1] College of Mechanical Engineering, Hangzhou Dianzi University, Hangzhou 310018, China; 212010040@hdu.edu.cn (C.L.); 211010028@hdu.edu.cn (G.W.); 01003@hdu.edu.cn (G.C.); 222010130@hdu.edu.cn (Y.G.); student_zdq@163.com (D.Z.); 41439@hdu.edu.cn (C.C.)
[2] Anji Intelligent Manufacturing Technology Research Institute Co., Ltd., Hangzhou Dianzi University, Huzhou 313000, China
[3] Sanmen Sanyou Technology, Inc., Taizhou 317103, China; limz_ja@163.com
[4] Changzhou Slav Intelligent Equipment Technology Co., Ltd., Changzhou 213162, China; 42335@hdu.edu.cn
* Correspondence: wwenxindiaolong@hdu.edu.cn

Abstract: This study aims to develop equipment for the preparation of composite films and successfully implement a film thickness prediction function. During the research process, we segmented the mechanical structure of the composite thin film preparation equipment into distinct modules, completed the structural design of the core module, and validated the stability of the process chamber, as well as the reasonableness of the strength and stiffness through simulation. Additionally, we devised a regression model for predicting the film thickness of composite films. The input features for the model included the sputtering air pressure, sputtering current, and sputtering time for magnetron sputtering process samples, as well as the evaporation volume and evaporation current for vacuum evaporation process samples. Simultaneously, the output features were the film thickness for both process samples. Subsequently, we established the designed composite film preparation equipment and conducted experimental verification. During the experiments, we successfully prepared Cr-Al composite films and utilized AFM for surface morphology analysis. The results confirmed the excellent performance of the Cr-Al composite films produced by the equipment, demonstrating the reliability of the equipment.

Keywords: composite film; preparation platform; machine learning; Cr-Al; film thickness prediction

Citation: Li, C.; Li, W.; Wu, G.; Chen, G.; Wu, J.; Zhang, N.; Gan, Y.; Zhang, D.; Chen, C. Design and Study of Composite Film Preparation Platform. *Crystals* **2024**, *14*, 389. https://doi.org/10.3390/cryst14050389

Academic Editor: Anelia Kakanakova

Received: 3 April 2024
Revised: 17 April 2024
Accepted: 18 April 2024
Published: 23 April 2024

1. Introduction

Composite films, comprising alternating layers of materials with outstanding properties, find widespread applications in various fields. The intricate preparation process demands efficient, stable, controllable equipment and precise film thickness control, posing a crucial challenge in composite film production [1,2].

Cr-Al composite film is a multifunctional material with wide and diverse applications. These films play an important role in the following fields. (1) Solar cell technology: Cr-Al composite films are used as the back passivation layer of solar cells to improve the conversion efficiency of the battery; (2) high-barrier packaging: these films are used in food and medicine and high-barrier packaging of electronic products, extending product life and improving food safety; (3) mechanical engineering: Cr-Al composite films are widely used in tools, molds, and mechanical parts due to their high strength, hardness, and wear resistance; (4) medical devices, optical films, and flat panel displays: these films are also used in other areas such as medical devices, optical coatings, and flat panel displays.

Amidst the rapid advancements in science and technology [3–5], the current design status and developmental trends of composite thin film preparation platforms have garnered significant attention [6–8]. In 2023, a foreign research team devised an integrated ultrahigh-vacuum cluster system to address interfacial spin effects in spintronic multilayer films [9,10]. Simultaneously, a Japanese research team proposed a gas-injected pulsed plasma CVD method utilizing a single plasma source as an ultra-plasma deposition technique for preparing DLC films with a nanoindentation hardness of 17.5 GPa. During the same period, Antonio A. A. Chepluki, Tiago E. A. Frizon, and others developed a low-cost spin-coater for thin film deposition, serving as a cost-effective alternative to high-priced commercial equipment [11–15].

This research delves into the intricacies of integrating and optimizing composite thin film preparation platforms [16]. By consolidating the preparation processes of magnetron sputtering and vacuum evaporation within the same vacuum cavity, the creation of composite films tailored to the requirements of diverse high-tech fields on a single substrate is achieved [17]. Furthermore, an integrated learning algorithm is introduced for film thickness prediction, facilitating accurate forecasts for the film thickness of composite films [18–20]. Finally, the paper substantiates the performance and feasibility of the equipment by preparing Cr-Al composite films on PET substrate. These innovative optimizations present an effective solution to the design and prediction challenges faced by hybrid film preparation systems [21–23].

2. Materials and Methods

2.1. Hardware Design

2.1.1. Overall Design

The mechanical structure design of the composite film preparation equipment outlined in this paper primarily encompasses a vacuum system capable of achieving a vacuum, a vacuum chamber creating a vacuum test environment, a process system integrating a magnetron sputtering system and a vacuum evaporation system, a transmission system responsible for opening and closing the vacuum chamber, and a cooling system for temperature control [24]. Combining the aforementioned analysis and the composition of each design component, the definitive overall structure of the composite film preparation equipment was established [25,26]. The comprehensive assembly diagram of the mechanical approach for the hybrid film preparation equipment is illustrated in Figure 1.

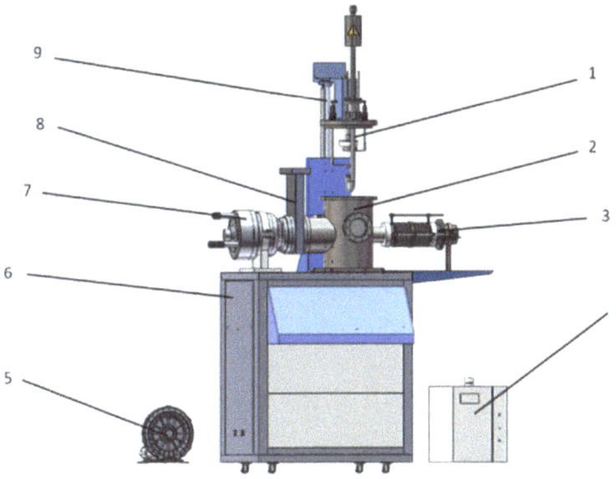

Figure 1. Assembly diagram of composite film preparation equipment. 1—sputtering system; 2—process chamber; 3—evaporation system; 4—water cooler; 5—vacuum pump; 6—electronic control cabinet; 7—molecular pump; 8—baffle valve; 9—transmission system.

2.1.2. Magnetron Sputtering Module Design

The sputtering system is primarily composed of a sputtering chamber, magnetron sputtering target, sample stage, sputtering target head baffle, sample stage baffle, target head holder, and other structures [27–29]. The structural schematic sketch and physical drawings are presented in Figures 2a and 2b, respectively.

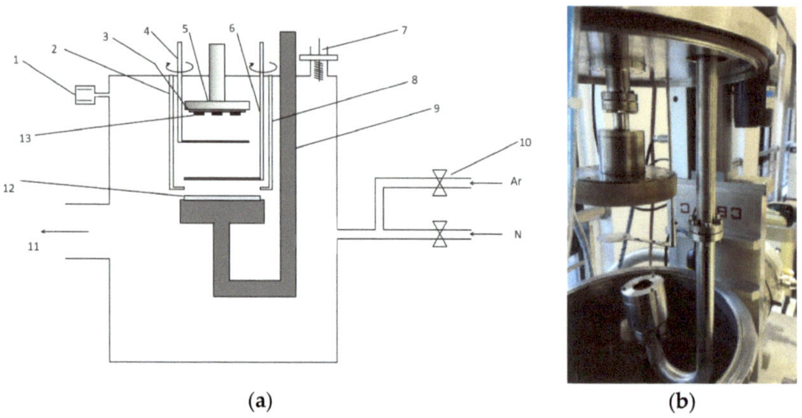

Figure 2. Schematic diagram of sputtering system. (**a**) Two-dimensional schematic; 1—capacitance manometer; 2—grounding shield; 3—substrate; 4—sample stage baffle; 5—height-adjustable substrate stage; 6—sputtering target head baffle; 7—ion piezometer; 8—grounding shield; 9—target head holder; 10—sputtering gas control valve; 11—connecting vacuum pump; 12—target material; 13—substrate heater; (**b**) physical image of the sputtering system.

2.1.3. Evaporation System Module Design

The vacuum evaporation module stands as a pivotal component of this preparation platform, harmoniously integrated with the magnetron sputtering module to form the composite film preparation platform's comprehensive process. The evaporation system comprises the process chamber, evaporation crucible, water-cooled pipeline, heating platform, heat source bellows, heat transfer components, and other integral elements. The structural schematic principle is illustrated in Figure 3 below.

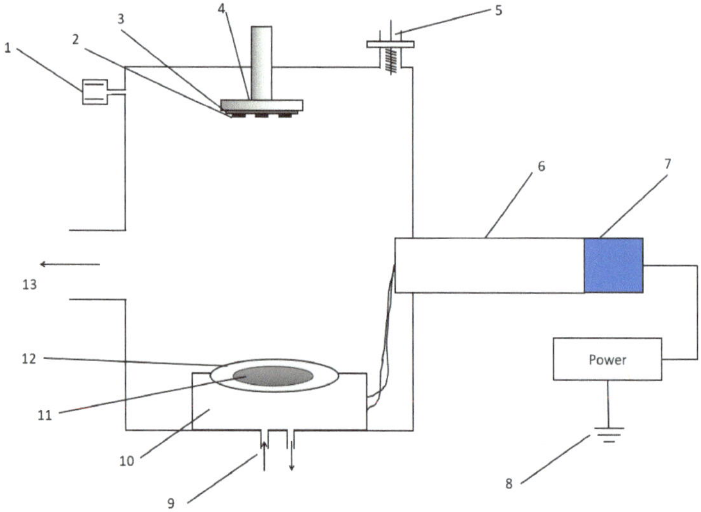

Figure 3. Evaporation system structure principle sketch: 1—capacitance manometer; 2—substrate heater; 3—substrate; 4—height-adjustable substrate table; 5—temperature gauge; 6—bellows heat transfer assembly; 7—heat source flange for heating platform; 8—ground shield; 9—water-cooled piping; 10—heating platform for evaporation system; 11—evaporation target; 12—evaporation crucible; 13—connection vacuum pump.

Within the evaporation system, the evaporation crucible assumes a central role, necessitating a thorough analysis of material selection and structural design. Considering that vacuum evaporation plating employs metal and organic molecular materials, operates within a temperature range of 200 to 1300 °C, and requires efficient cooling, alumina is chosen as the primary material for the evaporation crucible. Based on the relevant literature and information, the standardized specifications for the universal evaporation crucible in vacuum evaporation plating are an outer diameter of 19.6 mm, inner diameter of 15.4 mm, and height of 24 mm. A tungsten filament serves as the heat source, with the thermocouple positioned at the crucible's bottom for convenient temperature measurement and control. Standard configurations incorporate S-type thermocouples, and the evaporation working temperature range spans 200–1300 °C. The distance between the evaporation module and the sample stage is 60 mm. The schematic structure of the evaporation crucible is depicted in Figure 4 below, and the physical representation is presented in Figure 5.

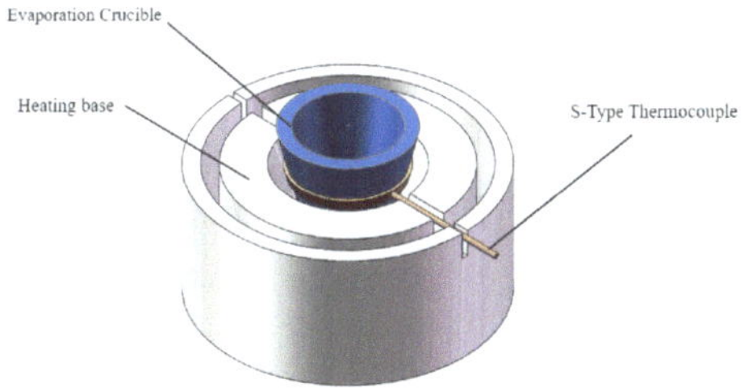

Figure 4. Structural design diagram.

Figure 5. Physical image.

2.1.4. Vacuum System Module Design and Simulation

In accordance with the force requirements of the vacuum chamber, materials possessing ample strength, stiffness, toughness, heat resistance, corrosion resistance, and other pertinent physical and chemical properties are selected. In domestic film-forming equipment manufacturing, 304 stainless steel is the chosen material.

The shape of the vacuum vessel is determined based on the characteristics of the substrate and deposition source, with cylindrical and rectangular designs being the primary choices. Cylindrical designs offer high strength and are well suited for small- to medium-sized vessels, while rectangular designs optimize space utilization. For film-forming equipment, smaller- to medium-sized equipment typically adopts a vertical cylindrical design, while larger equipment leans towards a horizontal cylindrical configuration. The composite film preparation equipment embraces a vertical cylinder design, with the cylinder welded to the bottom plate. The schematic representation of the designed vacuum chamber structure is depicted in Figure 6. The cylindrical vacuum chamber features an inner diameter of 250 mm and a height of 360 mm, and the wall thickness is 7 mm. It is equipped with CF100 flanges and a quartz window, a CF60 flange interface to connect the molecular pump, an LF250 flange interface for installing the sputtering target and the sample stage, and a KF63 flange interface for integrating the evaporation module Additionally, two adjustable needle valves at the rear control the air inlet, and a reserved 1/16-inch interface accommodates the mass spectrometer.

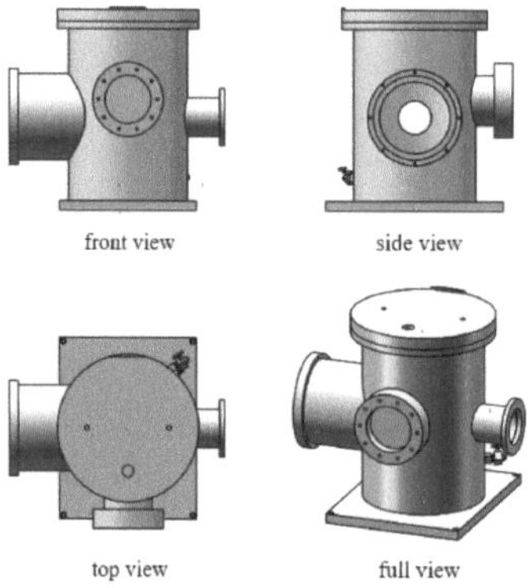

Figure 6. Vacuum chamber structure.

As the central space for experiments, a thorough analysis of the vacuum chamber's strength, stiffness, and stability through performance simulation is essential. In real operational conditions, the vacuum chamber maintains an internal pressure of 0.0001 Pa and is exposed to an atmospheric pressure of 0.1 Mpa externally. Under equivalent conditions, a negative pressure of −0.1 MPa is applied internally to simulate evacuation. The vacuum chamber, placed on the electric control cabinet without considering its self-weight undergoes structural static analysis. The results, illustrating the stress field distribution and deformation distribution, are presented in Figure 7. It can be seen from Figure 7a,b that under actual working conditions of the vacuum chamber, maximum deformation occurs at the center of the flange cover on the vacuum chamber, with the maximum value being 9.2926×10^{-3} mm. This value is very small and has almost no impact on the deformation and performance of the vacuum chamber. According to the simulation results in Figure 7b, the maximum deformation of the vacuum chamber is 9.29×10^{-3} mm. At this scale, the deformation of the vacuum chamber will not affect the overall equipment operation ecology, nor will it cause instability, so it meets the strain design requirements. Maximum stress occurs at the junction of the lower flange of the vacuum chamber cover and the inner wall of the vacuum chamber. The maximum value is 8.04 MPa, which is far less than the allowable stress of the material 150 MPa. Because this position is welded, maximum stress occurs at this position. It is also realistic, so the design is reasonable.

From Figure 8a,b, it is evident that under actual working conditions, maximum deformation occurs in the center of the upper flange cover, measuring 9.2926×10^{-3} mm This value is sufficiently small, having no impact on the performance and aligning with the design requirements of staying within 0.1 mm. Maximum stress is observed in the lower flange and the inner wall junction, registering at 8.04 MPa, significantly below the permissible stress of 150 MPa. This outcome aligns with the actual welding conditions affirming the reasonability of the design.

Under the specified conditions, an external air pressure of 10^5 Pa is applied to the vacuum chamber, and a thermodynamic coupling analysis simulation is executed to assess performance in vacuum and heated environments. In a vacuum setting, heat transfer occurs through contact position exchange and thermal radiation. Key parameters include a sputtering target heating source at 500 °C, substrate table heating at 100 °C, sputtering target and substrate table surface emissivity set at 0.8, and the vacuum cavity inner wall emissivity at 0.1. The temperature distribution results are then integrated into the static analysis module to complete the thermal coupling analysis. The obtained results for the stress field and deformation distribution are illustrated in Figure 8.

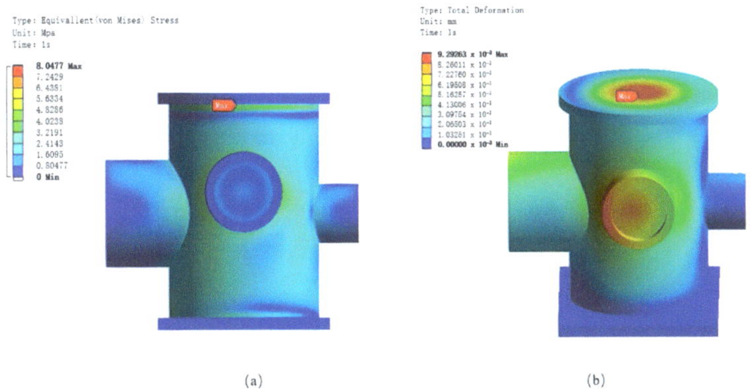

Figure 7. Static stress field distribution and deformation distribution resultant plots; (**a**) static stress field distribution; (**b**) static deformation field distribution.

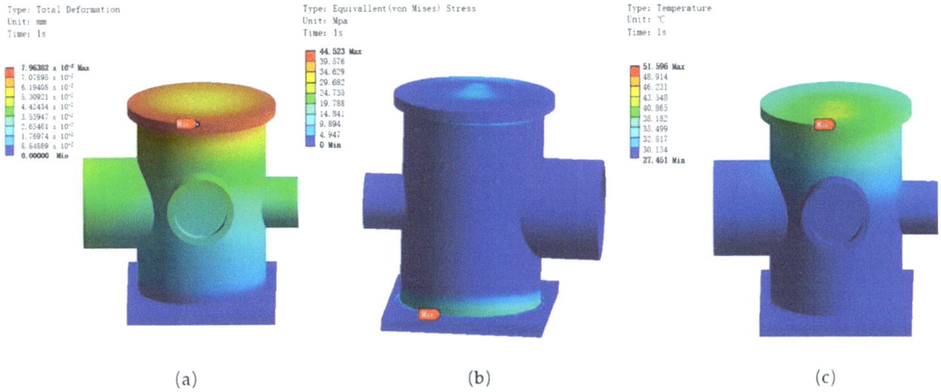

Figure 8. Thermal coupling analysis of the stress deformation results of the figure; (**a**) thermal coupling deformation field distribution; (**b**) thermal coupling strain field distribution; (**c**) temperature distribution map.

As depicted in Figure 8a,b, hot air radiation and sputtering target heat transfer affect the vacuum cover, resulting in a maximum deformation of 7.96×10^{-3} mm, well below the 0.1 mm threshold. The vacuum cavity remains unaffected, and maximum stress is observed in the vacuum cavity and the bottom plate weld, measuring 44.523 MPa. This value is significantly less than the permissible stress for 304 stainless steel, which is 150 MPa. Consequently, the structural design and material selection for the vacuum cavity align with the practical requirements.

2.2. Research on Film Thickness Prediction Algorithm

In the preceding chapters, our focus was on investigating the process and structural design of composite thin film preparation equipment, specifically addressing the magnetron sputtering and vacuum vapor deposition processes. This chapter focuses on achieving accurate predictions of the film thickness of composite films, delving into the design of a film thickness prediction model based on secondary integration learning. We utilized the PyCharm compiler software 2022.3.2 on the PC to execute the quadratic integration algorithm model, scripted in Python. Various algorithmic models were compared to validate the feasibility and advantages of the quadratic integration learning model.

2.2.1. Modeling of the Secondary Integration Learning Algorithm

The construction of a secondary integrated learning model involves the collection of input and output features, the creation of a dataset, its division into a training set and a test set, model training, and subsequent evaluation using performance metrics [30,31]. The specific process is depicted in Figure 9.

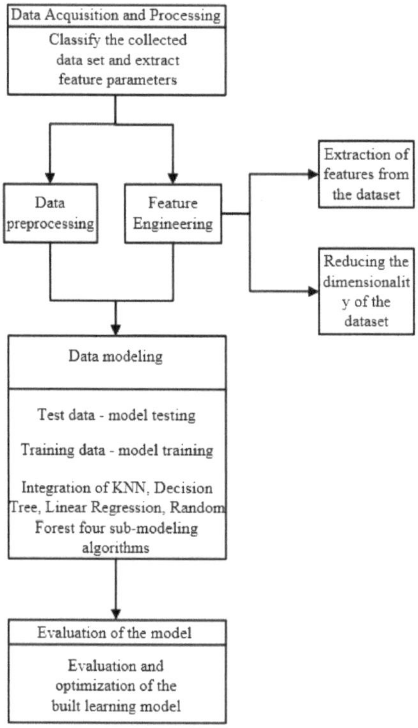

Figure 9. Secondary integration learning modeling process.

2.2.2. Feature Selection and Data Segmentation

In predicting the film thickness of the composite film, the characteristic inputs for the magnetron sputtering process include sputtering air pressure, current, and time, with the film thickness as the output. Similarly, for the vacuum evaporation process, the inputs encompass the evaporation amount and resistance evaporation current, yielding the film thickness as the output. The predicted film thicknesses from both approaches are aggregated to obtain the overall predicted film thickness of the composite film [32,33].

To enhance prediction accuracy, this article selects 15 samples for the magnetron sputtering process and 10 samples for the vacuum evaporation process. These samples were randomly divided into training and test sets, and the ratio of training and test sets was ensured to be 4:1.

Dataset division is a critical step. A standard method allocates four-fifths of the dataset for training and parameter optimization, reserving the remaining one-fifth for testing model performance. This approach ensures a relatively independent test set while retaining a substantial number of training sets, facilitating the accurate evaluation of model performance in real scenarios. The training and test datasets for both processes are presented in Tables 1 and 2, with an asterisk denoting the training set.

Table 1. Magnetron sputtering process dataset division: the data group with * is the training set used to train the model, and the data group without * is the test set used to evaluate the model.

Dataset No.	Sputtering Air Pressure/Pa	Sputtering Current/mA	Sputtering Time/s	Film Thickness/nm
1 *	0.7	120	10	10.6
2	0.8	120	14	10.4
3 *	1.4	120	30	16.0
4 *	1.5	120	14	10.8
5	0.9	120	12	11.2
6 *	0.8	120	20	11.5
7 *	0.7	130	10	12.3
8 *	0.8	130	14	11.9
9 *	1.4	130	30	17.5
10 *	1.5	130	14	13.3
11	0.8	110	20	10.5
12 *	0.9	120	13	11.1
13 *	0.8	120	18	10.8
14 *	0.8	120	30	19.3
15 *	0.9	120	35	25.4

Table 2. Vacuum vapor deposition process dataset division: the data group with * is the training set used to train the model, and the data group without * is the test set used to evaluate the model.

Dataset No.	Evaporation Amount/g	Resistance to Vaporization Current/A	Film Thickness/nm
1 *	0.2	30	0.2
2 *	0.4	30	0.3
3	0.6	30	0.4
4 *	0.8	30	0.6
5 *	1.0	25	0.7
6 *	1.0	30	0.7
7 *	1.0	35	0.7
8 *	1.0	40	0.8
9	1.2	30	0.8
10 *	1.5	30	1.1

2.2.3. Model Training and Effectiveness Evaluation

We constructed linear regression (LR), a random forest model (RFM), K-nearest neighbors (KNNs), decision tree (DT), and bagging and boosting models using Python in the PyCharm IDE. The performance of these models in terms of film thickness prediction was evaluated using metrics such as explained variance score (EV), mean absolute error (MAE), mean square error (MSE), mean absolute percentage error (MAPE), and the coefficient of determination R2 score [34–37].

We have selected 10 sets of experimental conditions for both magnetron sputtering and vacuum evaporation processes to compare the predicted(test) values from the models with the actual values. Figure 10 shows the comparison of the predicted results (Y_Pred) and the actual values (Y_true) obtained from 10 different experimental conditions in the magnetron sputtering process, using a linear regression model (a), decision tree model (b), random forest model (c), KNN model (d), bagging model (e), and boosting model (f). Similarly, Figure 11 shows the comparison in the vacuum evaporation process under 10 different experimental conditions using the same set of models [38,39]. This is applicable to Figures 10 and 11. In the picture, X-axis: number of samples; Y_True: film thickness true value; Y_Pred: film thickness prediction.

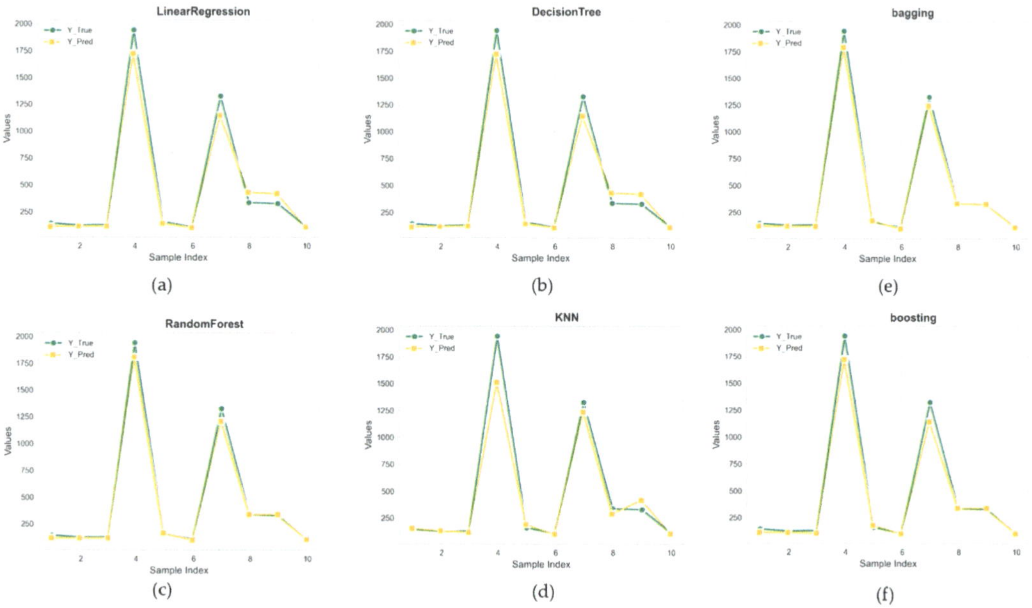

Figure 10. Comparison chart of the test results from six prediction models versus the actual results in the magnetron sputtering process. (**a**) Comparison chart of the Linear Regression model; (**b**) Comparison chart of the Decision Tree model; (**c**) Comparison chart of the Random Forest mode; (**d**) Comparison chart of the KNN model; (**e**) Comparison chart of the Bagging model; (**f**) Comparison chart of the Boosting model.

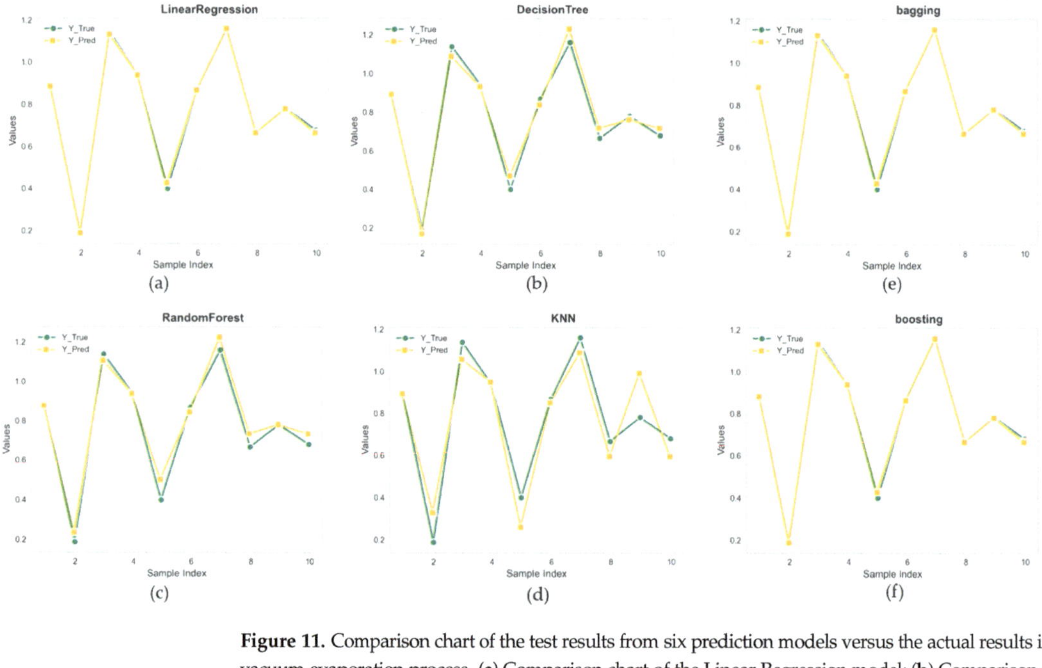

Figure 11. Comparison chart of the test results from six prediction models versus the actual results in the vacuum evaporation process. (**a**) Comparison chart of the Linear Regression model; (**b**) Comparison chart of the Decision Tree model; (**c**) Comparison chart of the Random Forest mode; (**d**) Comparison chart of the KNN model; (**e**) Comparison chart of the Bagging model; (**f**) Comparison chart of the Boosting model.

Figure 11 illustrates the training effects of the four sub-models for the vacuum vapor deposition process—decision tree, random forest, KNN, and linear regression—consistently with the magnetron sputtering process. At the same time, the training effect of the quadratic integral model of the vacuum evaporation process is also depicted. In the training effect plot, the horizontal and vertical axes signify the same parameters as in the magnetron sputtering process. In the picture, X-axis: number of samples; Y_True: film thickness true value; Y_Pred: film thickness prediction.

Both Figures 10 and 11 represent the comparison of the test results and actual results for each model in both the vacuum evaporation process and the magnetron sputtering process, revealing a close consistency between the predicted values and the actual values. Therefore, it is recommended to use the Bagging model and the Boosting model for secondary ensemble learning.

3. Results

3.1. Platform Hardware Introduction

To validate the effectiveness of the selected equipment for composite film preparation, experimental validation was conducted. The choice of equipment was guided by considerations of applicability, precision, and adaptability to various process conditions. Drawing upon the preceding design and selection criteria, the experimental platform depicted in Figure 12 was constructed. This platform encompasses essential components such as the vacuum module, magnetron sputtering module, vacuum evaporation module, and transmission lifting module.

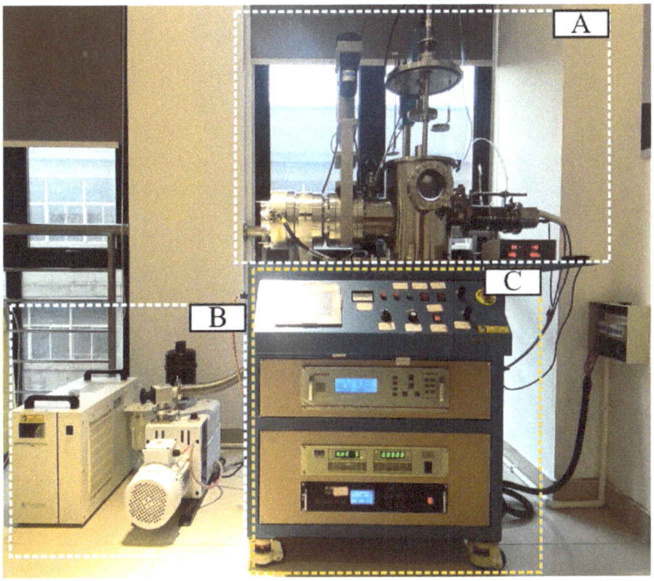

Figure 12. Experimental platform of composite film preparation equipment. (**A**) Mechanical structure of hybrid film preparation equipment; (**B**) execution equipment of hybrid film preparation equipment; (**C**) composite film preparation equipment.

3.2. Surface Topography Inspection

The 2D–3D diagrams of Cr films, Cr-Al composite films, and Al-Cr composite films prepared by composite film preparation equipment are shown in Figure 13.

The 2D–3D morphology of the Cr substrate film, prepared through magnetron sputtering, is depicted in Figure 13a,b, with its maximum thickness measured at 0.22 μm.

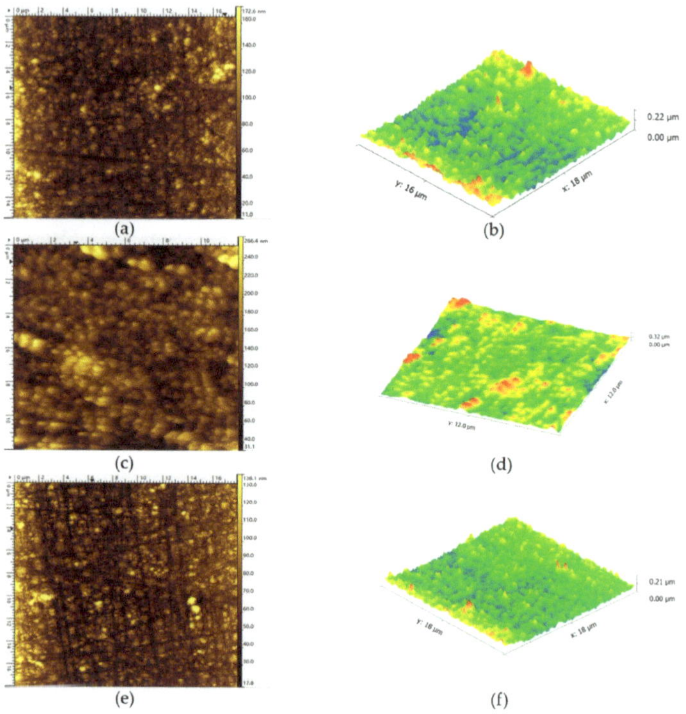

Figure 13. (**a**) Two-dimensional morphology of Cr bottom membrane; (**b**) three-dimensional morphology of Cr base film; (**c**) two-dimensional morphology of Cr-Al composite film; (**d**) three-dimensional morphology of Cr-Al composite film; (**e**) two-dimensional morphology of Al-Cr composite film; (**f**) three-dimensional morphology of Al-Cr composite film.

Subsequently, an Al film was deposited on this Cr base film through vacuum vapor deposition, resulting in the formation of the Cr-Al composite film. The 2D–3D diagrams of this composite film are presented in Figure 13c,d, while the integrated diagrams illustrating the texture, waveform, and roughness of the Cr-Al composite film are shown in Figure 14. The analysis of Figures 13c,d and 14 reveals that the thickness of the Cr-Al composite film reaches a maximum of 0.32 μm. Further examination of the texture, waveform, and roughness of the Cr-Al composite film indicates that the Al film, serving as the upper layer, exhibits the advantages of a uniformly distributed and relatively smooth surface.

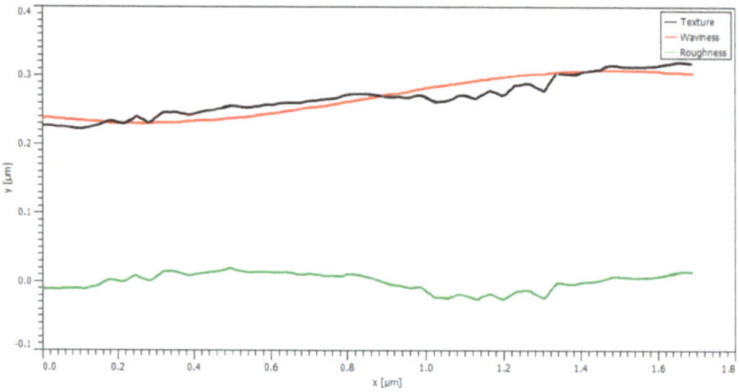

Figure 14. Combined texture, waveform, and roughness of Cr-Al composite film; x-axis: film expansion length; y-axis: texture, waviness, and roughness change value.

The deposition method we used involves magnetron sputtering for the underlying Cr film and vacuum evaporation for the top Al film. This method has been proven to form uniform films over a large area. Our experimental results show that the thickness variation of the Cr-Al film we formed across the entire sample is less than 100 nm, demonstrating its good uniformity.

The same materials are used in different sequences to prepare Al-Cr composite films. The 2D and 3D diagrams and roughness and other parameter distribution diagrams are shown in Figures 13e,f and 15.

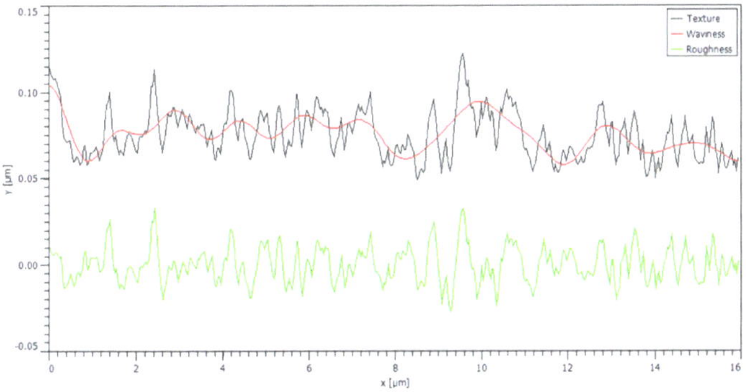

Figure 15. Combined texture, waveform, and roughness of Al-Cr composite film; x-axis: film expansion length; y-axis: texture, waviness, and roughness change value.

As evidenced by Figures 13e,f and 15, the peak of the Al-Cr composite film is 0.21 µm, with the entire film thickness fluctuating within the 0~0.21 µm range. The comprehensive diagram of the texture, waveform, and roughness of the Al-Cr composite film indicates a significant variation in roughness, which is attributed to vibrations or warping during the magnetron sputtering process. The texture and waveform in the diagram also exhibit considerable fluctuations. In conjunction with the two-dimensional and three-dimensional morphology of the Al-Cr composite film, it can be inferred that the presence of large-sized particles on the film surface, i.e., regional non-uniformity, causes this. The deepening of the grooves between large grains also leads to an increase in roughness. This analysis reveals that the magnetron sputtering process exhibits small-area non-uniformity and large-area uniformity, which may be caused by the inherent vibrations or warping in the magnetron sputtering process.

3.3. Algorithm Model Validation

Upon analyzing the results predicted by the established secondary integrated learning algorithmic model, the evaluation metrics for the magnetron sputtering process and vacuum evaporation process are presented in Tables 3 and 4 below.

Table 3. Metrics for evaluating the regression model of a magnetron sputtering process.

Assessment of Indicators	LR	KNN	DT	RF	Bagging	Boosting
EV	0.91	0.95	0.97	0.99	0.99	0.98
MAE	146.46	72.94	65.39	32.01	31.87	51.15
MSE	33,237.01	20,529.71	9951.06	3173.00	3035.97	8242.24
MAPE	59.44	12.22	12.86	6.42	7.01	9.98
R2	0.91	0.94	0.97	0.99	0.99	0.98

Table 4. Regression model evaluation metrics for the vacuum evaporation process.

Assessment of Indicators	LR	KNN	DT	RF	Bagging	Boosting
EV	1.00	0.87	0.98	0.98	1.00	1.00
MAE	0.01	0.08	0.04	0.04	0.01	0.01
MSE	0.00	0.01	0.00	0.00	0.00	0.00
MAPE	1.15	17.77	5.84	8.28	1.15	1.11
R2	1.00	0.87	0.98	0.97	1.00	1.00

The evaluation indexes for the magnetron sputtering process prediction model (Table 3) reveal that the comparative ranking of the training model results for the magnetron sputtering process is as follows: bagging > DF > boosting > DT > KNN > LR. Meanwhile, for the vacuum evaporation process training model (Table 4), the comparative ranking of the results is as follows: bagging $\geq$ LR > boosting > DT > RF > KNN. Notably, the regression model fit value of the integrated learning bagging model achieves 100%, signifying a perfect model. When comparing the different regression models for both preparation processes, it is evident that in each approach, our designed secondary integrated learning bagging model demonstrates optimal performance, reaching 99% in the magnetron sputtering process and achieving a perfect 100% fit in the vacuum evaporation process. This suggests that the robust regression model designed with secondary integrated learning bagging surpasses ordinary weak machine algorithm models, substantiating its superior performance.

4. Discussion

This study undertook a comprehensive scheme design for the structure of composite film preparation equipment, successfully implementing the film thickness prediction function. Throughout the research process, the following key achievements were realized: (1) The mechanical structure of the composite thin film preparation equipment was systematically divided into different modules, completing the structural design of the core module. Simulation verification ensured the stability and reasonable strength stiffness of the process chamber. (2) The design and implementation of a regression model for predicting the film thickness of composite thin films were achieved. Input features included the sputtering air pressure, sputtering current, and sputtering time for magnetron sputtering process samples, as well as the evaporation amount and evaporation current for vacuum evaporation process samples. The film thickness served as the output feature for both process samples. (3) Experimental validation was conducted using the designed composite film preparation equipment. Cr-Al composite thin films were successfully prepared, and AFM surface morphology analysis confirmed the excellent performance and reliability of both the equipment and the Cr-Al composite films prepared.

Through the comparative analysis of the results of the Cr-Al composite film and the Al-Cr composite film, it is known that there is a step phenomenon on the surface of the single-layer Cr film sputtered by the magnetron in this device. When a composite film is prepared using vacuum evaporation, the step phenomenon will also exist, which is caused by the step-type bottom layer film. By comparing the three-dimensional morphology and height distribution of the two, it can be concluded that the Cr-Al film prepared by first magnetron sputtering and then vacuum evaporation by this device has better uniformity. The magnetron sputtered Cr film serves as the bottom layer, giving the Cr-Al composite film excellent step performance. Furthermore, the Al film prepared by vacuum evaporation as the upper functional layer has better surface uniformity and smoothness. This will allow for the performance of the composite film to be maximized, thereby enhancing its performance, providing this film with better application scenarios and fields.

In the future, the design of composite film preparation platforms should trend towards the greater integration of multiple processes to facilitate the preparation of more complex composite films. Additionally, a focus on enhanced prediction accuracy is recommended, exploring advanced techniques such as deep learning or neural networks to make the models more adaptable to dynamic and evolving preparation conditions.

Author Contributions: Conceptualization, C.L. and W.L.; methodology, C.L.; software, Y.G.; validation, G.C., W.L. and C.C.; formal analysis, J.W. investigation, N.Z.; resources, G.C.; data curation, C.L.; writing—original draft preparation, C.L.; writing—review and editing, C.L.; visualization, W.L.; supervision, W.L.; project administration, G.C.; funding acquisition, D.Z. and G.W. All authors have read and agreed to the published version of the manuscript.

Funding: This research was supported by Natural Science Foundation of Zhejiang Province, China (No. ZCLQ24E0101), Zhejiang Science and Technology Plan Project (No. 2022C01199), and Zhejiang Science and Technology Plan Project (No. 2023C01065).

Data Availability Statement: The original contributions presented in the study are included in the article, further inquiries can be directed to the corresponding author/s.

Conflicts of Interest: Author Niushan Zhang was employed by the company Changzhou Slav Intelligent Equipment Technology. Junyi Wu was employed by the company Sanmen Sanyou Technology. The remaining authors declare that the research was conducted in the absence of any commercial or financial relationships that could be construed as a potential conflict of interest.

References

1. Abdelaziz, B.B.; Mustapha, N.; Bedja, I.M.; Aldaghri, O.; Idriss, H.; Ibrahem, M.; Ibnaouf, K.H. Spectral Behavior of a Conjugated Polymer MDMO-PPV Doped with ZnO Nanoparticles: Thin Films. *Nanomaterials* **2023**, *13*, 2405. [CrossRef]
2. Behl, C.; Behlert, R.; Seiler, J.; Helke, C.; Shaporin, A.; Hiller, K. Characterization of Thin AlN/Ag/AlN-Reflector Stacks on Glass Substrates for MEMS Applications. *Micro* **2024**, *4*, 142–156. [CrossRef]
3. Alanazi, T.I. Design and Device Numerical Analysis of Lead-Free $Cs_2AgBiBr_6$ Double Perovskite Solar Cell. *Crystals* **2023**, *13*, 267. [CrossRef]
4. Ali, D.O.A.; Fabbiani, M.; Coulomb, L.; Bosc, S.; Villeroy, B.; Estournès, C.; Estournès, C.; Koza, M.M.; Beaudhuin, M.; Viennois, R. Reactive Spark Plasma Sintering and Thermoelectric Properties of Zintl Semiconducting $Ca_{14}Si_{19}$ Compound. *Crystals* **2023**, *13*, 262. [CrossRef]
5. Gong, L.; Zhang, P.; Lou, Z.; Wei, Z.; Wu, Z.; Xu, J.; Chen, X.; Xu, W.; Wang, Y.; Gao, F. Effect of Bi^{3+} Doping on the Electronic Structure and Thermoelectric Properties of $(Sr_{0.889-x}La_{0.111}Bix)TiO_{2.963}$: First-Principles Calculations. *Crystals* **2023**, *13*, 178. [CrossRef]
6. Brito, D.; Anacleto, P.; Pérez-Rodríguez, A.; Fonseca, J.; Santos, P.; Alves, M.; Cavalli, A.; Sharma, D.; Claro, M.S.; Nicoara, N.; et al. Antimony Selenide Solar Cells Fabricated by Hybrid Reactive Magnetron Sputtering. *Nanomaterials* **2023**, *13*, 2257. [CrossRef]
7. Castillo, I.; Mishra, K.K.; Katiyar, R.S. Characterization of V_2O_3 Nanoscale Thin Films Prepared by DC Magnetron Sputtering Technique. *Coatings* **2022**, *12*, 649. [CrossRef]
8. Ding, Y.; Wang, Y.; Chen, J.; Chen, W.; Hu, A.; Shu, Y.; Zhao, M. Substrate-Assisted Laser-Induced Breakdown Spectroscopy Combined with Variable Selection and Extreme Learning Machine for Quantitative Determination of Fenthion in Soybean Oil. *Photonics* **2024**, *11*, 129. [CrossRef]
9. Stoddard, N.; Pimputkar, S. Progress in Ammonothermal Crystal Growth of Gallium Nitride from 2017–2023: Process, Defects and Devices. *Crystals* **2023**, *13*, 1004. [CrossRef]
10. Zhou, F.; Zhao, Y.; Fu, F.; Liu, L.; Luo, Z. Thickness Nanoarchitectonics with Edge-Enhanced Raman, Polarization Raman, Optoelectronic Properties of GaS Nanosheets Devices. *Crystals* **2023**, *13*, 1506. [CrossRef]
11. Estime, B.; Ren, D.; Sureshkumar, R. Tailored Fabrication of Plasmonic Film Light Filters for Enhanced Microalgal Growth and Biomass Composition. *Nanomaterials* **2023**, *14*, 44. [CrossRef] [PubMed]
12. Hwang, S.; Han, Y.; Gardner, D.J. Characterization of CNC Nanoparticles Prepared via Ultrasonic-Assisted Spray Drying and Their Application in Composite Films. *Nanomaterials* **2023**, *13*, 2928. [CrossRef] [PubMed]
13. Ji, S.; Zhu, J.; Yang, Y.; Zhang, H.; Zhang, Z.; Xia, Z.; Zhang, Z. Self-Attention-Augmented Generative Adversarial Networks for Data-Driven Modeling of Nanoscale Coating Manufacturing. *Micromachines* **2022**, *13*, 847. [CrossRef] [PubMed]
14. Junrear, J.; Sakunasinha, P.; Chiangga, S. The Optimization of Metal Nitride Coupled Plasmon Waveguide Resonance Sensors Using a Genetic Algorithm for Sensing the Thickness and Refractive Index of Diamond-like Carbon Thin Films. *Photonics* **2022**, *9*, 332. [CrossRef]
15. Kotlyar, V.; Nalimov, A.; Kovalev, A.; Stafeev, S. Optical Polarization Sensor Based on a Metalens. *Sensors* **2022**, *22*, 7870. [CrossRef] [PubMed]

16. Ku, C.-A.; Yu, C.-Y.; Hung, C.-W.; Chung, C.-K. Advances in the Fabrication of Nanoporous Anodic Aluminum Oxide and Its Applications to Sensors: A Review. *Nanomaterials* **2023**, *13*, 2853. [CrossRef]
17. Lee, M.; Kim, J.; Khine, M.T.; Kim, S.; Gandla, S. Facile Transfer of Spray-Coated Ultrathin AgNWs Composite onto the Skin for Electrophysiological Sensors. *Nanomaterials* **2023**, *13*, 2467. [CrossRef] [PubMed]
18. Lin, Q.; Wang, Z.; Meng, Q.; Mao, Q.; Xian, D.; Tian, B. A Co-Sputtering Process Optimization for the Preparation of FeGaB Alloy Magnetostrictive Thin Films. *Nanomaterials* **2023**, *13*, 2948. [CrossRef]
19. Ling, L.; Zhang, X.; Hu, X.; Fu, Y.; Yang, D.; Liang, E.; Chen, Y. Research on Spraying Quality Prediction Algorithm for Automated Robot Spraying Based on KHPO-ELM Neural Network. *Machines* **2024**, *12*, 100. [CrossRef]
20. Loghin, F.; Rivadeneyra, A.; Becherer, M.; Lugli, P.; Bobinger, M. A Facile and Efficient Protocol for Preparing Residual-Free Single-Walled Carbon Nanotube Films for Stable Sensing Applications. *Nanomaterials* **2019**, *9*, 471. [CrossRef]
21. Hashimoto, T.; Letts, E.R.; Key, D. Progress in Near-Equilibrium Ammonothermal (NEAT) Growth of GaN Substrates for GaN-on-GaN Semiconductor Devices. *Crystals* **2022**, *12*, 1085. [CrossRef]
22. Heng, C.; Wang, X.; Zhao, C.; Wu, G.; Lv, Y.; Wu, H.; Zhao, M.; Finstad, T.G. Ultrathin Rare-Earth-Doped MoS_2 Crystalline Films Prepared with Magnetron Sputtering and Ar + H_2 Post-Annealing. *Crystals* **2023**, *13*, 308. [CrossRef]
23. Luo, J.; Khattinejad, R.; Assari, A.; Tayyebi, M.; Hamawandi, B. Microstructure, Mechanical and Thermal Properties of Al/Cu/SiC Laminated Composites, Fabricated by the ARB and CARB Processes. *Crystals* **2023**, *13*, 354. [CrossRef]
24. Manjunath, M.; Hausner, S.; Heine, A.; De Baets, P.; Fauconnier, D. Electrical Impedance Spectroscopy for Precise Film Thickness Assessment in Line Contacts. *Lubricants* **2024**, *12*, 51. [CrossRef]
25. Martínez, C.; Arcos, C.; Briones, F.; Machado, I.; Sancy, M.; Bustamante, M. The Effect of Adding CeO_2 Nanoparticles to Cu–Ni–Al Alloy for High Temperatures Applications. *Nanomaterials* **2024**, *14*, 143. [CrossRef] [PubMed]
26. Michman, E.; Oded, M.; Shenhar, R. Dual Block Copolymer Morphologies in Ultrathin Films on Topographic Substrates: The Effect of Film Curvature. *Polymers* **2022**, *14*, 2377. [CrossRef] [PubMed]
27. Park, H.H.; Fermin, D.J. Recent Developments in Atomic Layer Deposition of Functional Overlayers in Perovskite Solar Cells. *Nanomaterials* **2023**, *13*, 3112. [CrossRef] [PubMed]
28. Redka, D.; Buttberg, M.; Franz, G. Chemical Vapor Deposition of Longitudinal Homogeneous Parylene Thin-Films inside Narrow Tubes. *Processes* **2022**, *10*, 1982. [CrossRef]
29. Wang, J.; Pan, Y.; Zhao, M.; Ma, P.; Lv, S.; Huang, Y. Computational Fluid Dynamics Numerical Simulation on Flow Behavior of Molten Slag–Metal Mixture over a Spinning Cup. *Processes* **2024**, *12*, 372. [CrossRef]
30. Zhang, Q.; Li, H.; Li, Y.; Wang, H.; Lu, K. A Dynamic Permeability Model in Shale Matrix after Hydraulic Fracturing: Considering Mineral and Pore Size Distribution, Dynamic Gas Entrapment and Variation in Poromechanics. *Processes* **2024**, *12*, 117. [CrossRef]
31. Zhou, W.-Y.; Chen, H.-F.; Tseng, X.-L.; Lo, H.-H.; Wang, P.J.; Jiang, M.-Y.; Fuh, Y.-K.; Li, T.T. Impact of Pulse Parameters of a DC Power Generator on the Microstructural and Mechanical Properties of Sputtered AlN Film with In-Situ OES Data Analysis. *Materials* **2023**, *16*, 3015. [CrossRef] [PubMed]
32. Calabretta, M.; Sitta, A.; Oliveri, S.M.; Sequenzia, G. Warpage Behavior on Silicon Semiconductor Device: The Impact of Thick Copper Metallization. *Appl. Sci.* **2021**, *11*, 5140. [CrossRef]
33. Grzywacz, H.; Jenczyk, P.; Milczarek, M.; Michałowski, M.; Jarząbek, D.M. Burger Model as the Best Option for Modeling of Viscoelastic Behavior of Resists for Nanoimprint Lithography. *Materials* **2021**, *14*, 6639. [CrossRef]
34. Wang, Y.-F.; Yoshida, J.; Takeda, Y.; Yoshida, A.; Kaneko, T.; Sekine, T.; Kumaki, D.; Tokito, S. Printed Composite Film with Microporous/Micropyramid Hybrid Conductive Architecture for Multifunctional Flexible Force Sensors. *Nanomaterials* **2023**, *14*, 63. [CrossRef]
35. Wu, R.; Hu, Y.; Li, P.; Peng, J.; Hu, J.; Yang, M.; Chen, D.; Guo, Y.; Zhang, Q.; Xie, X.; et al. Controlled Epitaxial Growth and Atomically Sharp Interface of Graphene/Ferromagnetic Heterostructure via Ambient Pressure Chemical Vapor Deposition. *Nanomaterials* **2021**, *11*, 3112. [CrossRef] [PubMed]
36. Yazdani, S.; Phillips, J.; Mosey, A.; Bsaibes, T.; Decca, R.; Cheng, R. Study of the Long-Range Exchange Coupling in Nd-Fe-B/Ti/Fe Multilayered Structure. *Crystals* **2024**, *14*, 119. [CrossRef]
37. Ying, M.; Liu, X.; Zhang, Y.; Zhang, C. Optimizing Load Capacity Predictions in Gas Foil Thrust Bearings: A Novel Full-Ramp Model. *Lubricants* **2024**, *12*, 76. [CrossRef]
38. Zhu, C.; Jin, L.; Li, W.; Han, S.; Yan, J. The Prediction of Wear Depth Based on Machine Learning Algorithms. *Lubricants* **2024**, *12*, 34. [CrossRef]
39. Zou, T.; Kang, L.; Zhang, D.; Li, J.; Zheng, Z.; Peng, X. Polyethylene Terephthalate Composite Films with Enhanced Flame Retardancy and Gas Barrier Properties via Self-Assembly Nanocoating. *Nanomaterials* **2023**, *13*, 2018. [CrossRef]

crystals

MDPI

Article

Growth and Properties of Ultra-Thin PTCDI-C8 Films on GaN(0001)

Katarzyna Lament [1], Miłosz Grodzicki [1,2], Radosław Wasielewski [1,*], Piotr Mazur [1] and Antoni Ciszewski [1]

1 Institute of Experimental Physics, University of Wrocław, pl. Maksa Borna 9, 50-204 Wrocław, Poland; katarzyna.lament2@uwr.edu.pl (K.L.); milosz.grodzicki@pwr.edu.pl (M.G.); antoni.ciszewski@uwr.edu.pl (A.C.)

2 Department of Semiconductor Materials Engineering, Faculty of Fundamental Problems of Technology, Wrocław University of Science and Technology, Wybrzeże Wyspiańskiego 27, 50-370 Wrocław, Poland

* Correspondence: radoslaw.wasielewski@uwr.edu.pl

Abstract: Ultra-thin PTCDI-C8 films are vapor-deposited under ultra-high vacuum (UHV) conditions onto surfaces of p- or n-doped GaN(0001) samples. The X-ray photoelectron spectroscopy (XPS) results reveal a lack of strong chemical interaction between the PTCDI-C8 molecule and the substrate. Changes in the electronic structure of the substrate or the adsorbed molecules due to adsorption are not noticed at the XPS spectra. Work function changes have been measured as a function of the film thickness. The position of the HOMO level for films of thicknesses 3.2–5.5 nm has been determined. Energy diagrams of the interface between p- and n-type GaN(0001) substates and the PTCDI-C8 films are proposed. The fundamental molecular building blocks of the PTCDI-C8 films on GaN(0001), assembled by self-organization, have been identified. They are rows of PTCDI-C8 molecules stacked in "stand-up" positions in reference to the substrate, supported by the π–π bonds which are formed between the molecular cores of the molecules and monomolecular layers constituted by rows which are tilted in reference to the layer plane. The layers are epitaxially oriented. The epitaxial relation between the rows and the crystallographic directions of the substrate are determined. A model of the PTCDI-C8 film's growth on GaN(0001) substrate is proposed. The 3D islands of PTCDI-C8 molecules formed on the substrate surface during film deposition are thermodynamically unstable. The Volmer–Weber type of growth observed here is a kinetic effect. Rewetting processes are noticeable after film aging at room temperature or annealing at up to 100 °C.

Keywords: thin films; organic layers; PTCDI-C8; semiconductors; GaN(0001)

Citation: Lament, K.; Grodzicki, M.; Wasielewski, R.; Mazur, P.; Ciszewski, A. Growth and Properties of Ultra-Thin PTCDI-C8 Films on GaN(0001). *Crystals* **2024**, *14*, 201. https://doi.org/10.3390/cryst14030201

Academic Editor: Alberto Girlando

Received: 22 January 2024
Revised: 12 February 2024
Accepted: 14 February 2024
Published: 20 February 2024

1. Introduction

There are two main factors determining the electric charge transport inside active layers of organic electronic devices: the molecular structure of the molecules constituting the layer and the supra-molecular organization of the layer. The first one depends on the current abilities of organic synthesis, which, until present, has achieved very high precision in adjusting the electronic properties of the molecules through their chemical modification [1–4]. The second one depends on interactions between the molecules and direct interactions between the substrate and the molecules. The charge transport in an organic semiconducting layer is directional. It prefers a characteristic direction. In the case of low-molecular-weight semiconductors, this is the direction of π stacking. Molecular disorder along this direction substantially lowers charge carrier mobility. It is therefore important from a practical point of view to produce optimally arranged organic films to minimize the negative effects of molecular disordering on the film's conductivity. There is also still very little known about the substrate's influence on the interactions responsible for the π stacking.

The N,N′-dioctyl-3,4:9,10-perylene tetracarboxylic diimide (PTCDI-C8) is one of the best n-type organic semiconductors currently available [5–8]. The electron mobility measured for organic thin-film transistors based on PTCDI-C8 equals up to ~1.7 cm^2/Vs [5] The molecule ($C_{40}H_{42}O_4N_2$) consists of a planar core and two alkyl chains on opposite sides. Its high molecular stability in air enables its operation in ambient conditions. The PCTDI-C8 molecules relatively easily organize themselves into supra-molecular architectures through hydrogen bonding, metal ion coordination and π stacking [9,10]. There are few studies on the properties of PTCDI-C8 thin films and they have been performed on various substrates such as SiO_2 and Al_2O_3; so far these molecules have not been studied on GaN(0001) [11,12].

GaN(0001) is the most frequently studied surface of gallium nitride (GaN). In combination with thin organic films, GaN surfaces offer several unique properties. The wide band gap of GaN allows optical access through the substrate and makes it easier to align the highest occupied and lowest unoccupied molecular orbitals of the organic film with the substrate band edges. It permits more flexibility in device design and, in the case of applications dependent on the charge transport across the interface, it increases the possibility for molecular control of the electronic properties of the hybrid organic–inorganic system [13–15] Due to its high electron mobility, chemical stability under physiological conditions, non-toxicity and biocompatibility, GaN is a very attractive material for biosensors [16–19]. It has been shown that by using the GaN thin-film high-electron-mobility transistors one can electrically detect proteins, antibodies, glucose and strands of DNA selectively and with high sensitivity [17,18].

The aim of the study reported herein is to characterize substrates' influence on the morphological, structural and electronic properties of PTCDI-C8 adsorption films on GaN(0001). The decisive factor for undertaking these experiments has been a willingness to identify substrate properties that have an influence on both the type of direct chemical bond between the molecules and the substrate and on the nature of the intermolecular bonds between the molecules inside the adsorption layer as well.

2. Experimental Details

The substrates used were 10 μm thick, (0001)-oriented, p-GaN (Mg-doped, 10^{18} cm^{-3}) and n-GaN (Si-doped, 10^{18} cm^{-3}) epitaxial layers deposited on Al_2O_3 (Technologies and Devices International, An Oxford Instruments Company, Oxford, UK). Typical size of the sample was about 4 × 8 mm^2. The samples were ex situ degreased in isopropanol and then washed in distilled water and dried in air. Before organic film evaporation, the substrate was in situ annealed at about 800 °C to remove any residual gases. This procedure allowed for the reduction of oxygen and carbon contaminations; however, they were not completely eliminated.

The samples were characterized in two separate UHV setups at room temperature (RT), using scanning tunneling microscopy (STM), X-ray photoelectron spectroscopy (XPS) ultraviolet photoelectron spectroscopy (UPS) and low-energy electron diffraction (LEED) techniques. The first setup included a VT STM/AFM microscope (Omicron,). The imaging was accomplished in the constant current mode using a tungsten tip. WSxM software(version number 5.0) was applied to analyze the STM results [20]. The XPS/UPS measurements were performed in the second UHV setup, equipped with Mg and Al anodes (Mg Kα (1253.6 eV) and Al Kα (1486.6 eV) lines) and a He I line (21.2 eV) radiation source. Due to the signals overlapping the Mg anode was used for the substrates N 1s and Ga 3d's lines measurements; the adsorbates C 1s and O 1s' lines were measured using the Al anode. Emitted photoelectrons were collected by a hemispherical electron energy analyzer (Phoibos 100-5, SPECS, Germany) with a pass energy of 10 or 2 eV and step size of 0.1 or 0.025 eV for core-level lines or a valance band, respectively. Optical axis of the analyzer entrance was normal to the substrate surface. The Fermi level position (E_F) was found by UPS measurement on a clean Au sample. The threshold of photoemission, which corresponds to the vacuum level of the sample, was also measured with a voltage (−5 V) applied to the samples to clear the detector's work function. The XPS spectra were analyzed using KolXPD (Kolibrik.net, Prague, Czech Republic) and/or CasaXPS software (version number 2.3.19 PR1.0). Deconvolution of the XPS peaks was modeled using Gaussian and Lorentzian line shapes and a Shirley-type background subtracting. LEED measurements were carried out in the energy range 0–300 eV with a step of 0.5 eV; diffraction patterns were recorded using a CCD camera.

Organic films of PTCDI–C8 molecules (98% purity, Sigma-Aldrich, MilliporeSigma, Burlington, MA, USA) were deposited by physical vapor deposition (PVD) on the substrates kept at room temperature (RT) under UHV, with a base pressure of ~10^{-10} Torr. The temperature of the quartz crucible used as the evaporator was about 300 °C. The evaporation rate at this temperature did not exceed 0.6 nm/min. The efficiency of the evaporator was calibrated by means of a quartz crystal resonator. The organic film growth and its characterization were performed step by step. In the case of XPS/UPS measurements, the average thickness of PTCDI-C8 films was additionally controlled on the bases of the Ga 3d substrate's line intensity decay (measured with the Mg anode) following the progress of adsorption layer growth, assuming a mean free path of electrons λ in PTCDI-C8 layer equal to 2.84 nm [21]. In the case of STM observations, the amount of deposited adsorbate was counted directly from the STM topographies.

3. Results

3.1. Samples Characterization Prior to Deposition

Prior to PTCDI-C8 deposition, the surface quality of the samples was evaluated by means of STM, LEED, XPS and UPS. Extended-area STM topographies of the p- and n-type GaN(0001) showed regularly stepped surfaces with a very small number of defects (Figure 1a,b). Their long-range atomic order was revealed by LEED (Figure 1c). Satisfactory STM imaging required relatively high bias voltages of around +5 V for p-type samples and about −5 V for n-type ones. The surfaces of the p- and n-GaN samples, subjected to the same cleaning procedures, did not differ in surface topography. The presence of terraces, which were tens of nanometers wide and a half or single GaN bi-layer high, were typical for both types of samples.

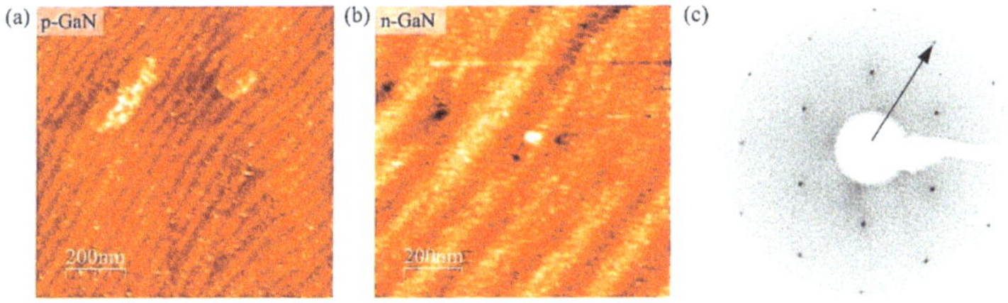

Figure 1. STM images of (**a**) p-GaN(0001) surface (imaged area equals to 1 × 1 μm^2, V_s = 5.1 V, I_t = 43.5 pA) and (**b**) n-GaN(0001) surface (1 × 1 μm^2, V_s = −5.1 V, I_t = 55.4 pA). Dark points correspond to the ends of screw dislocations. (**c**) Characteristic for both samples, the diffraction pattern exhibits hexagonal 1 × 1 structure with lattice constant of 0.319 nm (in this case the pattern is taken from an n-type sample by applying primary electron beam of energy equal to 150 eV). The arrow corresponds to <$2\bar{1}\bar{1}0$> direction in the real space.

XPS analysis of the samples revealed a surface oxygen concentration of approximately 20%. Unfortunately, carbon impurities could not be eliminated from the surfaces. The n-GaN surface exhibits a higher concentration of carbon contaminants compared to the p-GaN substrate. For both samples, the main N 1s and Ga 3d lines consisted of the same components. An example of the result of the deconvolution of these lines for p-GaN is presented in Figure 2. The Ga 3d line has four components (Figure 2a). The dominating one corresponds to Ga–N bonds. The component located at a higher binding energy relates to the presence of residual oxygen. The component denoted as Ga–Ga correlates to a metallic bond between gallium atoms [22]. Three components of the N 1s line are demonstrated in Figure 2b. The first one originates from Ga–N bonds, constituting the bulk of the line (denoted Ga–N (I) in Figure 2b); the second one comes from the surface Ga–N bonds; and the third one relates to nitrogen–hydrogen bonds (N–H_x) [23].

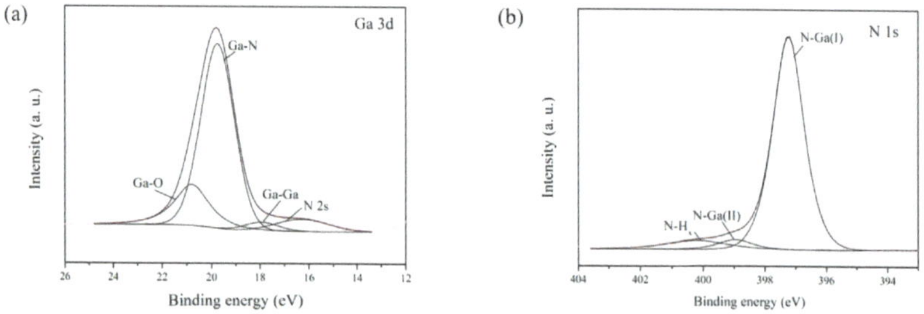

Figure 2. Components of (**a**) Ga 3d and (**b**) N 1s lines from p-GaN surface. See text for details.

The XPS analysis performed for n-GaN looks similar; however, the Ga 3d and N 1s lines are shifted by about 0.5 eV towards a higher binding energy. This comes from the different Fermi level locations on both p- and n-type surfaces [24].

The UPS spectra of the substrates, measured before PTCDI-C8 deposition, revealed the typical spectra for semiconductors. The position of the valence band maximum (VBM) can be determined by extrapolating the inclinations of the spectral curve in the region of the lowest binding energies (directly below the E_F). A predictable shift of the valence band edges between p- and n-type samples was observed. The VBM is at 1.7 eV and 3.1 eV for p- and n-type GaN, respectively. An example of the UPS spectrum for p-GaN is presented in Figure 3. The electron affinity of the GaN(0001) surface amounts to 4.0 eV and 3.3 eV, respectively, for p-GaN and n-GaN, as calculated from the relationship $\chi = hv - W - E_g$, where hv = 21.2 eV is the energy of photons; W is the width of the recorded spectrum, measured as the energy difference between the VBM and the cut-off threshold of the spectrum; and E_g = 3.4 eV is the GaN band gap width. The obtained results of the VBM's position and electron affinity are in line with other studies [22,25].

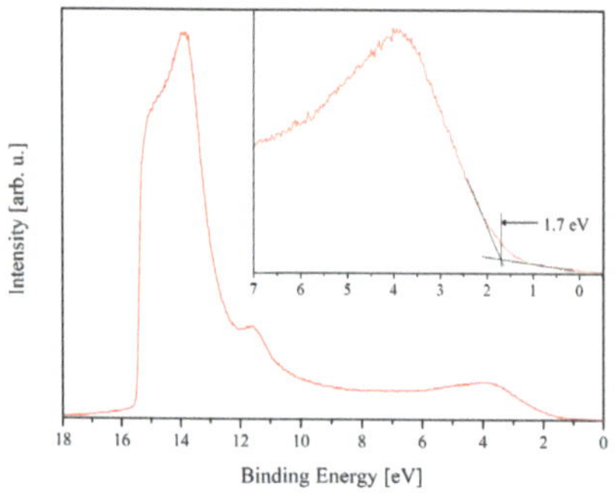

Figure 3. The UPS spectra measured for p-GaN(0001) samples prior to PTCDI-C8 deposition. Inserts depict the position of VBM.

3.2. The PTCDI-C8 Thin Films' Growth

The HOMO level signal becomes clearly visible in the UPS spectra when the thickness of the PCTDI-C8 film reaches 3.2 nm. An example of the UPS spectrum for the molecules on p-GaN is presented in Figure 4. The UPS results showed that the HOMO levels are located 1.9 eV and 2.2 eV below the E_F for the p- and n-GaN substrates, respectively.

Following the film's growth, changes in the work function ϕ, calculated from the expression $\phi = hv - E_{cut\text{-}off}$, where $E_{cut\text{-}off}$ is the cut-off threshold of the spectrum, follow different paths for p- and n-samples, as is seen in Figure 5. For p-GaN, ϕ keeps the value of 5.7 eV characteristic for the bare substrate up to an average film thickness d = 3.0 nm. From d = 3.2 nm onwards, the work function ϕ drops, reaching a value of 4.2 eV at an average thickness of 5.5 nm. In the case of n-GaN, the film growth brings about a work function increase after the first deposited dose of PCTDI-C8. The work function reaches 4.2 eV at an average thickness d = 1.8 nm.

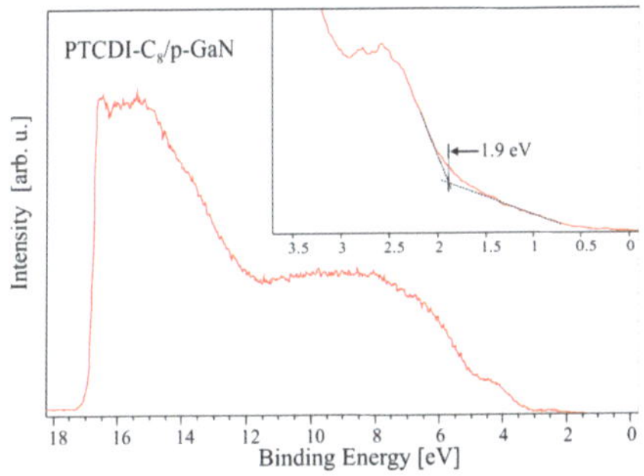

Figure 4. The UPS spectra taken for the 3.2 nm thick PTCDI-C8 film on p-GaN(0001). Inserts show the position of the leading edge of the HOMO level.

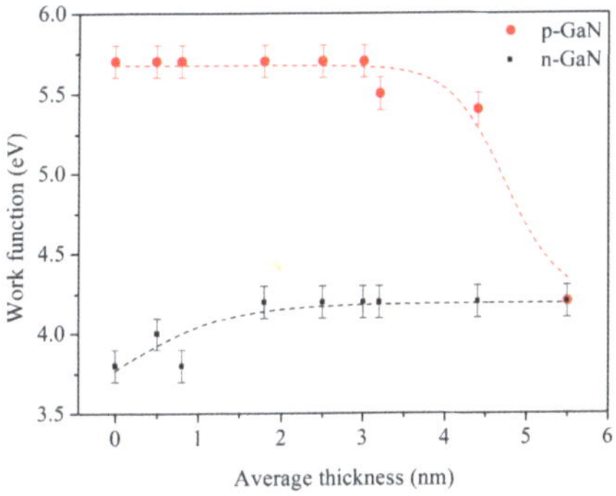

Figure 5. Work function changes as a function of the average thickness of the PTCDI-C8 layer on n- and p-GaN surfaces.

From the XPS spectra it is seen that PTCDI-C8 adsorption on a GaN(0001) surface does not alter the position of the dipper atomic levels of the atoms of both the surface and the molecule. Following the PTCDI-C8 film's growth, the main substrate lines Ga 3d and N 1s only lose their intensity, and do not undergo any shift. In Figure 6a, the C 1s lines of a bare p-GaN surface (spectrum (1)) and a surface covered with a 5.5 nm thick PTCDI-C8 film (spectrum (2)) are compared. For the bare substrate, the line contains only one peak (maximum at 284.6 eV) which originated from the remnants of carbon left on the surface after rapid annealing at the end of the cleaning procedure. In the case of the surface covered by the film, the line contains two components coming from the imide group (maximum at 288.0 eV) and molecule core (maximum at 285.0 eV), which are clearly identifiable. Both components increase their intensity following the film's growth. The N 1s line measured for the p-type sample before and after the 5.5 nm thick PTCDI-C8 film's deposition is shown in Figure 6b. For the bare substrate (spectrum (1)), the line consists of only one component associated with the Ga–N bonds (maximum at 397.3 eV). The height of the signal is scaled down ten times to fit the figure. After PTCDI-C8 deposition (spectrum (2)), in addition to this peak, attenuated by the deposited film substrate signal coming from the Ga–N bonds, another component appears in the spectrum of this line (with its peak at 400.2 eV) which originates from the imide group. These two species differ by 2.9 eV. The same energy difference between these two signals is also measured for the n-type sample. When the

Ga–N bond signal decreases following film growth, the signal originating from the imide group increases.

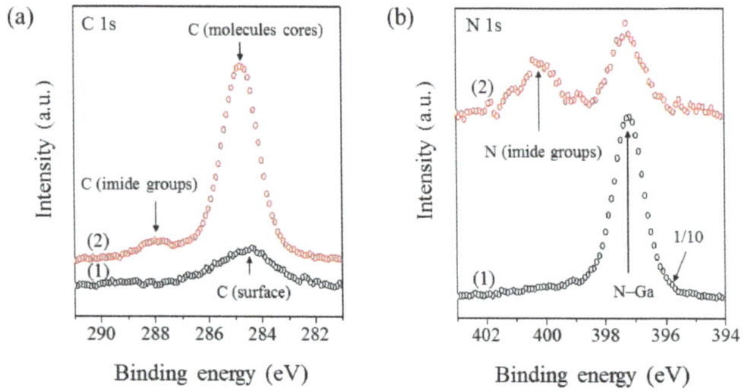

Figure 6. The XPS spectra of (**a**) the C 1s line, taken using Al anode, and (**b**) the N 1s line, taken using Mg anode. Spectra from bare p-GaN samples are denoted by (1). Spectra collected after deposition of the 5.5 nm thick PTCDI-C8 film are denoted by (2). See text for details.

Below a certain coverage of the substrate surface by PTCDI-C8 molecules, the STM images were of very low quality. In the case of p-GaN substantial improvement was reached when the average thickness of the deposited film exceeded 0.4 nm and formation of islands of adsorbate began. The STM topography of the substrate surface at this stage is shown in Figure 7a. The average height of the islands is about 1.5 nm (Figure 7b). The STM topography of part of the PTCDI-C8 island in Figure 7c reveals details of the shape of its edges. The magnified pattern of the area denoted by the dotted square in Figure 7c is in view in Figure 7d, revealing rows of molecules constituting an island. The distance between the two closest rows is equal to 1 nm and the rows are parallel to the [$2\bar{1}\bar{1}0$] direction of the substrate.

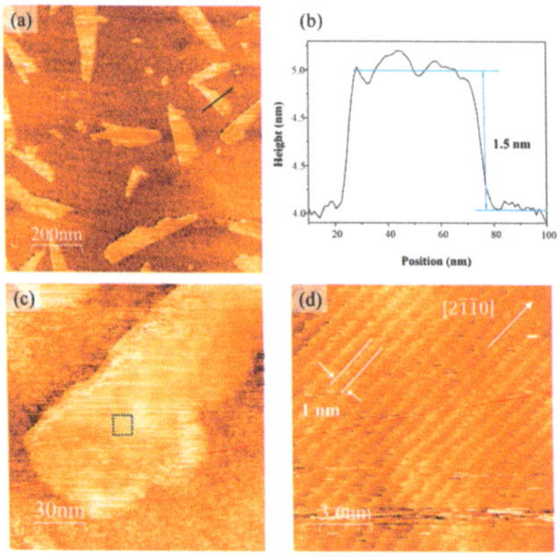

Figure 7. (**a**) The STM topography of two-dimentional molecular islands of PTCDI-C8 grown on p-GaN(0001) surface. The average thickness of the PTCDI-C8 film, estimated from the topography, is 0.4 nm. The islands are formed directly on the 1 × 1 μm^2 area of the substrate (imaging conditions V_s = 4.7 V, I_t = 107 pA). (**b**) Profile of the black line in (**a**). (**c**) Fragment of one of the islands and its surroundings, covering a 150 × 150 nm^2 area (V_s = 4.7 V, I_t = 182 pA); (**d**) Magnified part of the PTCDI-C8 island surface marked by a dotted square in (**c**) (18 × 18 nm^2, V_s = 4.7 V, I_t = 103 pA) revealing the rows of molecules constituting the island. Distance between the rows is equal to 1 nm and the rows are parallel to the [$2\bar{1}\bar{1}0$] direction of the substrate.

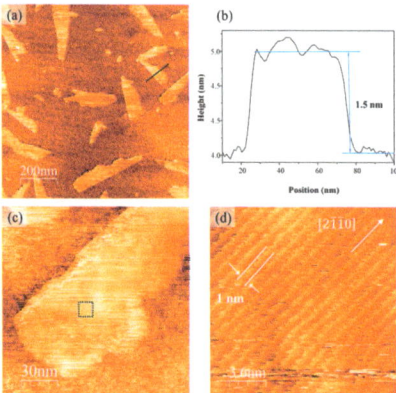

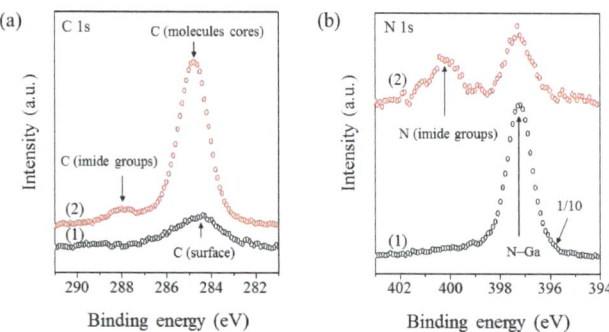

The results of using 2D fast Fourier transform (FFT) on the STM pattern of the ordered monomolecular layer of PTCDI-C8 on the p-GaN(0001) surface are shown in Figure 8. The FFT of the unfiltered STM pattern from Figure 8a is displayed in Figure 8b. The inverse Fourier transform obtained after noise filtering is seen in Figure 8c, exhibiting a long-range order along the rows as well. The distance between the two closest molecules in the row is equal to 0.6 nm. The islands of the first monomolecular layer deposited directly on the substrate surface are epitaxially oriented in relation to the substrate, with molecular rows mostly parallel to the densely packed $<\bar{1}\bar{1}20>$ or less frequently to the $<10\bar{1}0>$ substrate directions.

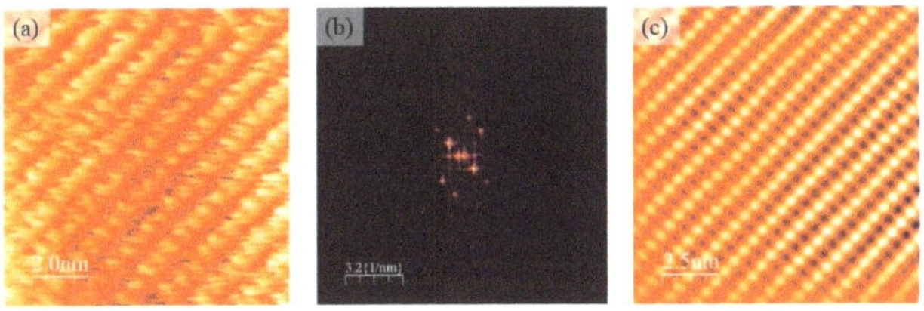

Figure 8. (**a**) The STM pattern of ordered monomolecular layer of PTCDI-C8 on p-GaN(0001) surface (10×10 nm^2, V_s = 4.7 V, I_t = 51 pA). (**b**) 2D FFT of the pattern. (**c**) Inverse FTT after noise filtering.

Under the deposition conditions applied here, the second monomolecular layer begins to grow before the first one is complete. An example of a PTCDI-C8 film composed of two monomolecular layers is shown in Figure 9. The total amount of deposited adsorbate corresponds to an average film thickness d = 1.8 nm. The first monomolecular layer of the topography in Figure 9a covers about 75% of the substrate surface. The second monomolecular layer covers about 17% of the surface of the first one. The profile of the superimposed layers is shown in Figure 9b. The height of each layer is equal to 2.0 nm. The edge of the upper layer is seen in the bottom right-hand corner of the topography in Figure 9c. The molecular rows of each layer are parallel, retaining the same distance between the rows of 1.2 nm. Due to coalescence, the islands of different rows' orientations form domains. The bottom PTCDI-C8 layer, visible on the right side of the topography in Figure 9d, consists of two domains A and B, who's rows are perpendicular. In this case, the rows of the upper layer are parallel to the rows of domain A. Nucleation of the third layer is noticed when the first layer covers 96% of the substrate and the second covers 70% of the first one. At this stage of the film growth, the heights of the layers amount to 2 nm in the case of the first layer and 1.7 nm in the case the second and the third layers.

The bilayer or three-layered PTCDI-C8 islands deposited on the substrate at RT under the conditions of our experiment are not thermally stable. After aging at RT (one hour or more), under UHV or short annealing (a few seconds) at up to 100 °C, the molecules from the upper layers diffuse down, completing the first layer and causing the growth and coalescence of islands in the first layer. As a result, the first layer, which has direct contact with the substrate, completes itself to the extent allowed for by the amount of deposited adsorbate. The STM topography of 3D islands constituting a PTCDI-C8 film of about 4.0 nm average thickness, just after deposition, is shown in Figure 10a. Some of the islands consist of three monomolecular layers, forming terraces that are very well distinguished from the terraces of the substrate (see Figure 1a,b). The topography in Figure 10b exhibits a PTCDI-C8 film of an average thickness of 1.0 nm after one hour of aging at RT. The film, which just after deposition was a bilayer film with a topography like that shown in Figure 9a, after aging, has been transformed into a monolayer film. The islands constituting the film are rounded and mostly merged, forming meandering chains. The same effect can be achieved after a few seconds of annealing. The topography in Figure 10c demonstrates a film of an average thickness of 1.5 nm. About 85% of the substrate surface is covered with the monolayer film, which, before the annealing, was a bilayer one. Similar behavior was also observed for films up to 5.5 nm thick.

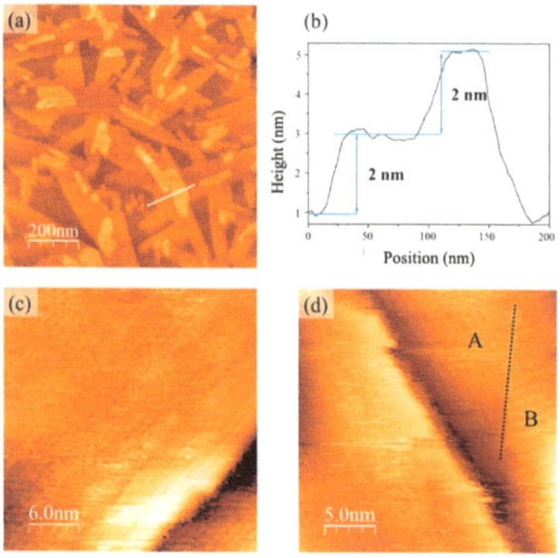

Figure 9. Morphology of the PTCDI-C8 film grown on a p-GaN(0001) surface at the stage of growth where the first layer of the molecules is not fully completed and the second layer is already growing. (**a**) Topography of a 1 × 1 µm^2 area of the sample. A profile of two superimposed layers measured along the white segment is shown in (**b**). (**c**) Topography of a 30 × 30 nm^2 area, revealing parallel molecular rows of the upper layer and the bottom layer, which are visible in the bottom right-hand corner of the pattern. Molecular rows of each layer retain the same distance of 1.2 nm. (**d**) Topography of a 25 × 25 nm^2 area in which the domain morphology of the bottom layer is visible. Dotted line divides domains A and B, who's rows are perpendicular.

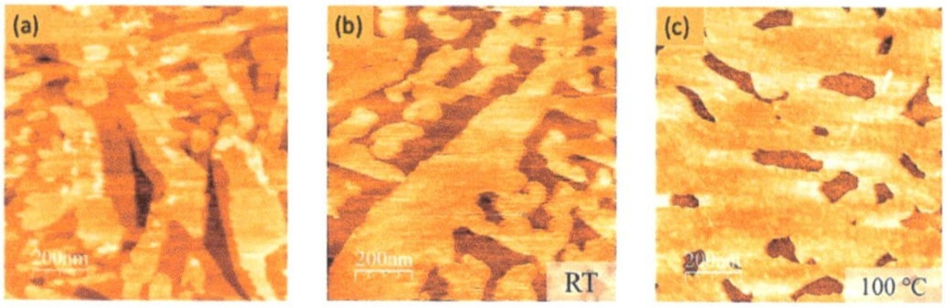

Figure 10. (**a**) The STM topography of 3D islands constituting a PTCDI-C8 film of about 4.0 nm average thickness, just after deposition. Some of the islands consist of three monomolecular layers. (**b**,**c**) The STM topographies of single-layer PTCDI-C8 films obtained from bilayer ones through rewetting mechanisms: (**b**) a bilayer film of an average thickness of 1.0 nm after aging for 1 h at RT, and (**c**) a bilayer film of an average thickness of 1.5 nm after annealing at up to 100 °C for a few seconds.

The initial growth stages of PTCDI-C8 films on n-GaN follow the same path as those for p-GaN. The only difference is the quality of the STM imaging. The STM patterns of the films growing on n-GaN are blurred up to a film thickness of about 1.8 nm, which corresponds to the stage at which the first PTCDI-C8 monolayer covers 70% of the substrate and the second layer start to grow, in opposition to p-GaN; where, for films with an average thickness exceeding 0.4 nm, imaging become substantially improved, allowing the detection of the morphological details of the growing film. The STM studies of thicker films on n-type substrates have been given up on at this stage.

4. Discussion

The STM topographies clearly show that carbon and oxygen residues, detected by XPS after applying the same cleaning procedures, were not uniformly distributed over the substrate surface. Instead, they were concentrated in groups, leaving an extended area of the substrate clean, with a characteristic terrace topography.

It is obvious that n- and p-type samples essentially differ in their electron structure. p-GaN(0001) has a work function ϕ = 5.7 eV and electron affinity χ = 4.0 eV. The n-type sample has lower values for its work function and electron affinity, respectively, 3.6 eV and 3.3 eV. The important features of the electron band structure of the PTCDI-C8/GaN(0001) interfaces are collected in the diagrams in Figure 11a,b. The E_F, in bulk, is located 3.3 eV and 0.1 eV above the valence band maximum (E_V) for the n- and the p-type samples, respectively [23]. The surface band bending at the vacuum/GaN(0001) interface for bare substrates, as calculated from the equation $V_C = (E_F - E_V)_{bulk} - (E_F - E_V)_{surface}$, is equal to 0.2 eV for the n-type and 1.6 eV for the p-type sample. The bending comes from electrostatic surface charging. Solving Poisson's equation, the space-charge region width $x_d = \left(\frac{2\varepsilon\varepsilon_0 V_C}{qN_a}\right)^{-\frac{1}{2}}$ can be obtained, where $\varepsilon = 8.9$ [25] is the dielectric constant of GaN, ε_0 represents the permittivity of free space and q is the elementary charge of an electron. The depletion layer width amounts to about 14 nm for n-GaN and about 40 nm in the case of p-GaN. Corrections connected to surface photovoltage (SPV) effects are not considered in the above estimations. It is known that, in the case of GaN, the SPV due to UPS or XPS radiation is determined to be ~0.5 V in magnitude [24]. In contrary to n-GaN, the band bending of p-GaN is strong due to the considerable depletion of holes in the near-surface region, caused by surface states which originate from Ga dangling bonds [26,27], thus, the Fermi level is pinned to these states, situating itself in the middle of the energy band gap [28–30].

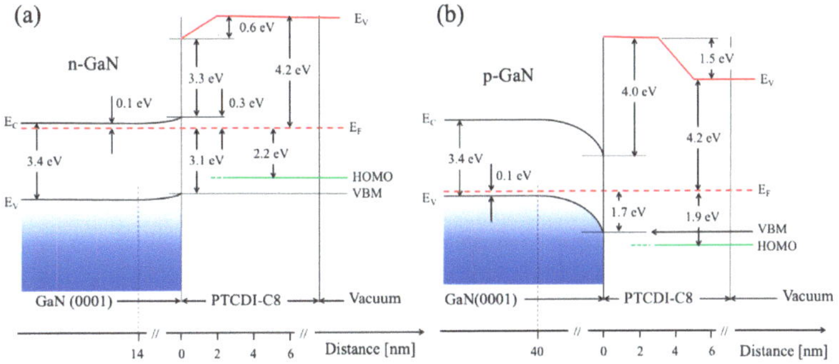

Figure 11. Energy band diagrams of the PTCDI-C8/GaN(0001) interface formed by a 5.5 nm thick film on: (**a**) n-doped and (**b**) p-doped GaN substrates. The left sides of each diagram correspond to the bulk band structure and show bending in the region near the interface. The right sides schematically illustrate the position of the vacuum level E_V as a function of the PTCDI-C8 layer thickness and the position of the HOMO level of the film in reference to its Fermi level.

From the XPS measurements it is seen that the interaction between the substrate and PTCDI-C8 film is very weak, and rather of a Van der Walls character. A stronger chemical interaction can be excluded because the Ga 3d and N 1s lines of the substrate do not change their positions or shapes following the first and successive doses of PTCDI-C8 deposition. Also, the C 1s lines originating from the imide group or from the core of the PTCDI-C8 molecule (Figure 6a), as well as the N 1s line from the imide group (Figure 6b), do not change their positions or shapes following the film growth. Taking these into account, it can be safely assumed that the influence of the PTCDI-C8 film on the band bending of the substrate can be neglected.

The 5.5 nm thick films have the same work function value, 4.2 eV, independent of the substrate type (n- or p-) onto which the film was deposited. The only difference in the electronic structure of the PTCDI-C8 films concerns their HOMO level position. In the case of the film on p-GaN, its level is situated 0.3 eV closer to the Fermi level than for the film on n-GaN. This shift of the HOMO level of the PTCDI-C8 film on p-GaN is caused, mos probably, by the same surface states which are responsible for strong band bending at th subsurface region of p-GaN.

There were no positive results of our STM studies regarding the adsorption of singl PTCDI-C8 molecules on the GaN(0001) surface. STM topographies of the surface afte the first dose of deposition are blurred, indicating that the adsorbate is weakly bounde and the molecules are mobile. In the case of p-GaN, the situation improves when th self-organization of the adsorbate begins and the two-dimensional islands start to grov In the case of n-GaN, the poor quality of the STM images made STM observations mor difficult at least up to the 1.8 nm thick films; the thickest PTCDI-C8 films on n-GaN studied herein using STM. It seems that at the first stages of the growth interaction betwee PTCDI-C8 molecules and the substrate are weaker on the n-type substrate. This could b caused by stronger rewetting mechanisms, which compete with the 3D growth mechanism Rewetting could be strengthened by the larger amount of carbon residue left on the n-typ surface after cleaning procedures. The presence of carbon on the surface is the only facto which chemically differentiates the n- and p-GaN substrates.

The self-organization of the PTCDI-C8 molecules deposited on p-GaN(0001) begin when the film reaches an average thickness of 0.4 nm, at which the first 2D islands ar observed. The islands have an ordered row structure. The analysis of the STM patterns o the films, for this range of thickness, reveals that the molecules in the row are in stand-u positions, as evidenced by the island's height of 1.5 nm (the thinnest observed), with a linear molecule packing density of 1.7×10^7 molecules/cm. This corresponds well with th arrangement of PTCDI-C8 molecules shown in Figure 12a. Although the dominant facto in self-organization is the interaction between molecules, the substrate structure influence the orientation of the island. The rows are parallel to the closely packed crystallographi directions of the substrate, in this case to [$2\bar{1}\bar{1}0$]. This means that the island's growth i epitaxial. The structure of the layers constituting the thin films on GaN(0001) surface proposed here is schematically shown in Figure 12. The molecules set in the way shown i Figure 12a interact through π bonds formed between the cores of the molecules creating th row. Tilted rows stacked one beside another compose the layer as it is shown in Figure 12b The height of the layer depends on the tilting angel of the rows. For the layer depicted i Figure 12b, the tilting angle is equal to ~30°, which makes the layer 1.5 nm high, with th distances between the alkyl tiles of the molecules forming a row equal to 1.0 nm, just lik for the PTCDI-C8 layer shown in Figure 9.

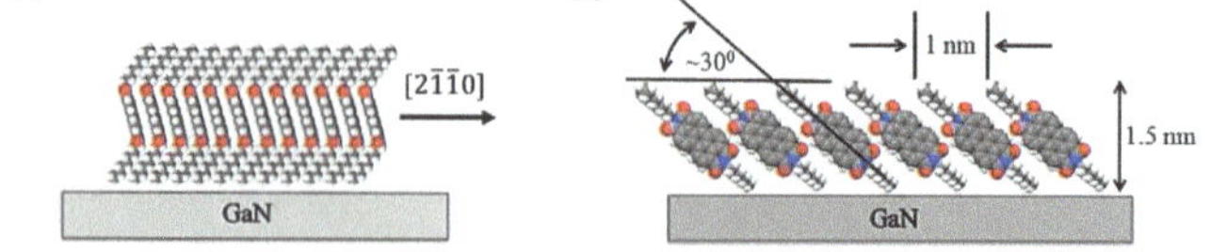

Figure 12. Schematically sketched structure of the first monomolecular layer of PTCDI-C8 film grown on a p-GaN(0001) surface: (**a**) view from the side of the row, positioned along the [$2\bar{1}\bar{1}0$] direction of the substrate, (**b**) view along the tilted rows which constitute the layer (or along the GaN [$2\bar{1}\bar{1}0$] direction).

The growth of 2D islands transforms into 3D growth when the average thickness o the film exceeds 1.2 nm. PTCDI-C8 molecules do not form a wetting layer. The substrate is not fully covered by the first layer when the second starts to grow. The increase in the average thickness d of the film results in the growth of consecutive layers of adsorbate or top of the bilayer. Three-layered islands bordering with uncovered by the adsorbate areas of the substrate are observed even for the 5.5 nm thick films. The molecular structure of the second and consecutive monomolecular layers is like the one sketched in Figure 12 It seems that height of the monomolecular layer increases following the increase in the quantity of the molecules composing the layer. Usually, the bottom layers of the island are thicker than the top one. This may be caused by a change in the tilting angel of the rows or the angle between the core and the alkyl tiles of the molecule. Similar mechanisms of growth and similar results concerning the morphology of the growing films of PTCDI-C8 and PTCDI-C13 were observed on SiO_2 surfaces [11,31].

The PTCDI-C8 films on GaN(0001) grow following the Volmer–Weber growth mode. The rewetting mechanisms observed in the PTCDI-C8/GaN(0001) system relax the thermodynamic instabilities in the morphology of the growing or already-grown films. The formation of instabilities is a kinetic effect that depends on growth parameters such as the flux of the deposited molecules or substrate temperature during deposition, as well as the intra- and interlayer surface diffusion controlled by Ehrlich–Schwöbel barriers (ESB). All these factors significantly influence the islands' nucleation and morphology evolution under conditions far from thermodynamic equilibrium [32]. The fundamental molecular building blocks of the PTCDI-C8 films on GaN(0001), produced by self-organization, are molecular rows formed by π–π bonds between the cores of the molecules and the layers assembled into rows by Van der Waals forces. Interactions between the layers constituting the 3D islands is also of a Van der Waals type. Under the growth condition used in this study during film deposition, the supersaturation of the 2D gas of the organic molecules adsorbed on the surface favors the formation of the critical nuclei of monomolecular layers on top of the already existing PTCDI-C8 layers, and as a consequence of the growth of the 3D terraced mounds. The terraces of the PTCDI-C8 mounds are clearly distinguishable from the terraces of the substrate (compare Figure 1a or Figure 1b with Figure 10a). The rewetting observed during the aging at RT or annealing at 100 °C results from the decay of the top-most layers, which are smaller, in the favor the lower ones, which are larger; this is the so-called "Ostwald ripening" [33]. The rewetting also shows that the ESB is lower for the diffusion of the organic molecule down-step of the terraces of the grown organic mounds than for the diffusion up-step. As observed here, Volmer–Weber growth is a kinetic effect; it seems possible to find such growth conditions as those at which the organic films could grow layer-by-layer, according to the Frank van der Merve growth mode.

5. Conclusions

The chemical composition and electronic structure of the surface and subsurface region, as well as the atomic structure and morphology of the bare n- and p-type GaN(0001) samples were characterized prior to PTCDI-C8 film deposition using XPS, UPS, LEED and STM—the differences between both types of the surfaces used as substrates have been discussed. We did not notice at the measured XPS spectra, any variations in the electronic structure of the substrate or adsorbed molecules due to PTCDI-C8 film growth; therefore, it has been assumed that the film–substrate interaction is of a Van der Waals character. Work function changes have been measured using UPS as a function of the average film thickness. The UPS measurements have allowed us to determine the position of the HOMO level for thicker films with an average thickness 3.2–5.5 nm. Energy diagrams of the interface between the n- and p-type GaN(0001) substates and the PTCDI-C8 films have been proposed. On the basis of STM observations, the fundamental molecular building blocks of the PTCDI-C8 films on GaN(0001), assembled by self-organization, have been identified. The first type of such blocks are rows of PTCDI-C8 molecules stacked in a "stand-up" position in reference to the substrate, supported by the π–π bonds which are formed between the molecular cores of the molecules. The second type are monomolecular layers constituted by rows which are tilted in reference to the layer plane. The layers are epitaxially oriented. The epitaxial relationship between the rows and the crystallographic directions of the substrate has been determined. Assuming that the interaction between the rows and between the layers is also of a Van der Waals forces origin, a model of the PTCDI-C8 film's growth on the GaN(0001) substrate is presented. The 3D islands of PTCDI-C8 molecules formed on the substrate surface during film deposition are thermodynamically unstable. The Volmer–Weber type of growth observed here is a kinetic effect. Rewetting processes are noticeable after film aging at room temperature or annealing at up to 100 °C.

Anisotropy and the complexity of the interactions inside the PTCDI-C8/GaN(0001) system produce a wide spectrum of basic phenomena, which are of great importance for understanding the growth of organic films on the surfaces of inorganic semiconductors. But there are also practical aspects to such studies. Elements of the organic film's morphology, such as the structure of the molecular building blocks of the film, its texture, its epitaxial relationship toward the substrate, etc., are key factors that determine its performance in various applications. A proper tuning of the growth parameters allows us to control all these elements; therefore, they may be used as tools for the engineering of hybrid organic–inorganic electronic devices.

Author Contributions: Conceptualization: A.C.; formal analysis: K.L., M.G. and R.W.; investigation: P.M.; writing—original draft preparation: A.C.; writing—review and editing: M.G. and R.W.; visualization: K.L., M.G., A.C. and R.W.; supervision: A.C. All authors have read and agreed to the published version of the manuscript.

Funding: This research received no external funding.

Data Availability Statement: The original data presented in the study are included in the article, further inquiries can be directed to the corresponding authors.

Conflicts of Interest: The authors declare no conflicts of interest.

References

1. Anthony, J.E.; Facchetti, A.; Heeney, M.; Marder, S.R.; Zhan, X. N-Type Organic Semiconductors in Organic Electronics. *Adv. Mater.* **2010**, *22*, 3876–3892. [CrossRef] [PubMed]
2. Zhan, X.; Facchetti, A.; Barlow, S.; Marks, T.J.; Ratner, M.A.; Wasielewski, M.R.; Marder, S.R. Rylene and Related Diimides for Organic Electronics. *Adv. Mater.* **2011**, *23*, 268–284. [CrossRef]
3. Pron, A.; Reghu, R.R.; Rybakiewicz, R.; Cybulski, H.; Djurado, D.; Grazulevicius, J.V.; Zagorska, M.; Kulszewicz-Bajer, I.; Verilhac, J.-M. Triarylamine Substituted Arylene Bisimides as Solution Processable Organic Semiconductors for Field Effect Transistors. Effect of Substituent Position on Their Spectroscopic, Electrochemical, Structural, and Electrical Transport Properties. *J. Phys. Chem. C* **2011**, *115*, 15008–15017. [CrossRef]
4. Gao, X.; Hu, Y. Development of N-Type Organic Semiconductors for Thin Film Transistors: A Viewpoint of Molecular Design. *J. Mater. Chem. C Mater.* **2014**, *2*, 3099–3117. [CrossRef]
5. Chesterfield, R.J.; McKeen, J.C.; Newman, C.R.; Ewbank, P.C.; Da Silva Filho, D.A.; Brédas, J.L.; Miller, L.L.; Mann, K.R.; Frisbie, C.D. Organic Thin Film Transistors Based on N-Alkyl Perylene Diimides: Charge Transport Kinetics as a Function of Gate Voltage and Temperature. *J. Phys. Chem. B* **2004**, *108*, 19281–19292. [CrossRef]
6. Gundlach, D.J.; Pernstich, K.P.; Wilckens, G.; Grüter, M.; Haas, S.; Batlogg, B. High Mobility N-Channel Organic Thin-Film Transistors and Complementary Inverters. *J. Appl. Phys.* **2005**, *98*, 064502. [CrossRef]
7. Tatemichi, S.; Ichikawa, M.; Koyama, T.; Taniguchi, Y. High Mobility N-Type Thin-Film Transistors Based on N,N′-Ditridecyl Perylene Diimide with Thermal Treatments. *Appl. Phys. Lett.* **2006**, *89*, 112108. [CrossRef]
8. Rolin, C.; Vasseur, K.; Schols, S.; Jouk, M.; Duhoux, G.; Müller, R.; Genoe, J.; Heremans, P. High Mobility Electron-Conducting Thin-Film Transistors by Organic Vapor Phase Deposition. *Appl. Phys. Lett.* **2008**, *93*, 033305. [CrossRef]
9. Würthner, F. Perylene Bisimide Dyes as Versatile Building Blocks for Functional Supramolecular Architectures. *Chem. Commun.* **2004**, *4*, 1564–1579. [CrossRef] [PubMed]
10. Pradhan, S.; Redwine, J.; McLeskey, J.T.; Dhar, A. Fabrication of N,N′-Dioctyl-3,4,9,10-Perylenedicarboximide Nanostructures through Solvent Influenced π–π Stacking and Their Morphological Impact on Photovoltaic Performance. *Thin Solid Films* **2014**, *562*, 423–429. [CrossRef]
11. Zykov, A.; Bommel, S.; Wolf, C.; Pithan, L.; Weber, C.; Beyer, P.; Santoro, G.; Rabe, J.P.; Kowarik, S. Diffusion and Nucleation in Multilayer Growth of PTCDI-C8 Studied with in Situ X-ray Growth Oscillations and Real-Time Small Angle X-ray Scattering. *J. Chem. Phys.* **2017**, *146*, 052803. [CrossRef]
12. Krauss, T.N.; Barrena, E.; Zhang, X.N.; De Oteyza, D.G.; Major, J.; Dehm, V.; Würthner, F.; Cavalcanti, L.P.; Dosch, H. Three-Dimensional Molecular Packing of Thin Organic Films of PTCDI-C 8 Determined by Surface X-ray Diffraction. *Langmuir* **2008**, *24*, 12742–12744. [CrossRef] [PubMed]
13. Seker, F.; Meeker, K.; Kuech, T.F.; Ellis, A.B. Surface Chemistry of Prototypical Bulk II–VI and III–V Semiconductors and Implications for Chemical Sensing. *Chem. Rev.* **2000**, *100*, 2505–2536. [CrossRef] [PubMed]
14. Gartsman, K.; Cahen, D.; Kadyshevitch, A.; Libman, J.; Moav, T.; Naaman, R.; Shanzer, A.; Umansky, V.; Vilan, A. Molecular Control of a GaAs Transistor. *Chem. Phys. Lett.* **1998**, *283*, 301–306. [CrossRef]
15. Cohen, R.; Kronik, L.; Shanzer, A.; Cahen, D.; Liu, A.; Rosenwaks, Y.; Lorenz, J.K.; Ellis, A.B. Molecular Control over Semiconductor Surface Electronic Properties: Dicarboxylic Acids on CdTe, CdSe, GaAs, and InP. *J. Am. Chem. Soc.* **1999**, *121*, 10545–10553. [CrossRef]

16. Jewett, S.A.; Makowski, M.S.; Andrews, B.; Manfra, M.J.; Ivanisevic, A. Gallium Nitride Is Biocompatible and Non-Toxic before and after Functionalization with Peptides. *Acta Biomater.* **2012**, *8*, 728–733. [CrossRef]
17. Pearton, S.J.; Ren, F.; Wang, Y.L.; Chu, B.H.; Chen, K.H.; Chang, C.Y.; Lim, W.; Lin, J.; Norton, D.P. Recent Advances in Wide Bandgap Semiconductor Biological and Gas Sensors. *Prog. Mater. Sci.* **2010**, *55*, 1–59. [CrossRef]
18. Kang, B.S.; Ren, F.; Wang, L.; Lofton, C.; Tan, W.W.; Pearton, S.J.; Dabiran, A.; Osinsky, A.; Chow, P.P. Electrical Detection of Immobilized Proteins with Ungated AlGaNGaN High-Electron-Mobility Transistors. *Appl. Phys. Lett.* **2005**, *87*, 023508. [CrossRef]
19. Baur, B.; Steinhoff, G.; Hernando, J.; Purrucker, O.; Tanaka, M.; Nickel, B.; Stutzmann, M.; Eickhoff, M. Chemical Functionalization of GaN and AlN Surfaces. *Appl. Phys. Lett.* **2005**, *87*, 263901. [CrossRef]
20. Horcas, I.; Fernández, R.; Gómez-Rodríguez, J.M.; Colchero, J.; Gómez-Herrero, J.; Baro, A.M. WSXM: A Software for Scanning Probe Microscopy and a Tool for Nanotechnology. *Rev. Sci. Instrum.* **2007**, *78*, 13705. [CrossRef]
21. Graber, T.; Forster, F.; Schöll, A.; Reinert, F. Experimental Determination of the Attenuation Length of Electrons in Organic Molecular Solids: The Example of PTCDA. *Surf. Sci.* **2011**, *605*, 878–882. [CrossRef]
22. Grodzicki, M. Properties of Bare and Thin-Film-Covered GaN(0001) Surfaces. *Coatings* **2021**, *11*, 145. [CrossRef]
23. Falta, J.; Schmidt, T.H.; Gangopadhyay, S.; Schulz, C.H.R.; Kuhr, S.; Berner, N.; Flege, J.I.; Pretorius, A.; Rosenauer, A.; Sebald, K.; et al. Cleaning and Growth Morphology of GaN and InGaN Surfaces. *Phys. Status Solidi (b)* **2011**, *248*, 1800–1809. [CrossRef]
24. Grodzicki, M.; Moszak, K.; Hommel, D.; Bell, G.R. Bistable Fermi Level Pinning and Surface Photovoltage in GaN. *Appl. Surf. Sci.* **2020**, *533*, 147416. [CrossRef]
25. Bougrov, V.; Levinshtein, M.E.; Rumyantsev, S.L.; Zubrilov, A. Galium Nitride (GaN). In *Properties of Advanced Semiconductor Materials GaN, AlN, InN, BN, SiC, SiGe*; Levinshtein, M.E., Rumyantsev, S.L., Shur, M.S., Eds.; John Wiley & Sons, Inc: Hoboken, NJ, USA, 2001; pp. 1–30.
26. Segev, D.; Van de Walle, C.G. Origins of Fermi-Level Pinning on GaN and InN Polar and Nonpolar Surfaces. *Europhys. Lett. (EPL)* **2006**, *76*, 305–311. [CrossRef]
27. Van De Walle, C.G.; Segev, D. Microscopic Origins of Surface States on Nitride Surfaces. *J. Appl. Phys.* **2007**, *101*, 081704. [CrossRef]
28. Long, J.P.; Bermudez, V.M. Band Bending and Photoemission-Induced Surface Photovoltages on Clean *n*- and *p*-GaN (0001) Surfaces. *Phys. Rev. B* **2002**, *66*, 121308. [CrossRef]
29. Grodzicki, M.; Mazur, P.; Ciszewski, A. Changes of Electronic Properties of P-GaN(0 0 0 1) Surface after Low-Energy N^+-Ion Bombardment. *Appl. Surf. Sci.* **2018**, *440*, 547–552. [CrossRef]
30. Wasielewski, R.; Mazur, P.; Grodzicki, M.; Ciszewski, A. TiO Thin Films on GaN(0001). *Phys. Status Solidi (b)* **2015**, *252*, 1001–1005. [CrossRef]
31. Vasseur, K.; Rolin, C.; Vandezande, S.; Temst, K.; Froyen, L.; Heremans, P. A Growth and Morphology Study of Organic Vapor Phase Deposited Perylene Diimide Thin Films for Transistor Applications. *J. Phys. Chem. C* **2010**, *114*, 2730–2737. [CrossRef]
32. Teichert, C.; Hlawacek, G.; Winkler, A.; Puschnig, P.; Draxl, C.; Teichert, C.; Hlawacek, G.; Winkler, A.; Puschnig, P.; Draxl, C. *Ehrlich-Schwoebel Barriers and Island Nucleation in Organic Thin-Film Growth*; Springer Series in Materials Science; Springer: Berlin/Heidelberg, Germany, 2013; Volume 173, pp. 79–106. [CrossRef]
33. Ratke, L.; Voorhees, P.W. *Growth and Coarsening*; Springer Science and Business Media: Dordrecht, The Netherlands, 2002. [CrossRef]

crystals

MDPI

Article

Sol–Gel Synthesis of ZnO:Li Thin Films: Impact of Annealing on Structural and Optical Properties

Tatyana Ivanova [1], Antoaneta Harizanova [1,*], Tatyana Koutzarova [2], Benedicte Vertruyen [3] and Raphael Closset [3]

1 Central Laboratory of Solar Energy and New Energy Sources, Bulgarian Academy of Sciences, Tzarigradsko Chaussee 72, 1784 Sofia, Bulgaria; tativan@phys.bas.bg
2 Institute of Electronics, Bulgarian Academy of Sciences, Tzarigradsko Chaussee 72, 1784 Sofia, Bulgaria; tatyana_koutzarova@yahoo.com
3 Group of Research in Energy and Environment from Materials (GREENMAT) Institute of Chemistry B6, University of Liege, B6aQuartier Agora, Allee du Six Août, 13, 4000 Liège, Belgium; b.vertruyen@ulg.ac.be (B.V.); raphael.closset@ulg.ac.be (R.C.)
* Correspondence: tonyhari@phys.bas.bg

Abstract: A sol–gel deposition approach was applied for obtaining nanostructured Li-doped ZnO thin films. ZnO:Li films were successfully spin-coated on quartz and silicon substrates. The evolution of their structural, vibrational, and optical properties with annealing temperature (300–600 °C) was studied by X-ray diffraction (XRD), Fourier Transform Infrared (FTIR), UV-VIS spectroscopic, and field emission scanning electron microscopic (FESEM) characterization techniques. It was found that lithium doping maintains the wurtzite arrangement of ZnO, with increasing crystallite sizes when increasing the annealing temperature. Analysis of the FTIR spectra revealed a broad main absorption band (around 404 cm^{-1}) for Li-doped films, implying the inclusion of Li into the ZnO lattice. The ZnO:Li films were transparent, with slightly decreased transmittance after the use of higher annealing temperatures. The porous network of undoped ZnO films was transformed to a denser, grained, packed structure, induced by lithium doping.

Keywords: sol–gel; ZnO; film coatings; doping; structural properties; optical transparency

Citation: Ivanova, T.; Harizanova, A.; Koutzarova, T.; Vertruyen, B.; Closset, R. Sol–Gel Synthesis of ZnO:Li Thin Films: Impact of Annealing on Structural and Optical Properties. *Crystals* **2024**, *14*, 6. https://doi.org/10.3390/cryst14010006

Academic Editors: Damian Wojcieszak, Michał Mazur and Miłosz Grodzicki

Received: 29 November 2023
Revised: 15 December 2023
Accepted: 15 December 2023
Published: 20 December 2023

1. Introduction

Zinc oxide (ZnO) has been shown to be a remarkable material with interesting physical and chemical properties [1]. ZnO is a wideband semiconductor, possessing a direct optical band gap (E_g) of 3.37 eV at room temperature, a large exciton binding energy (60 meV), and good transparency in the visible spectral range [2]. It is known to be a non-toxic, low-cost material with high radiation hardness and high thermal conductivity, while exhibiting a strong non-linear optical behavior [3]. Due to these promising properties, ZnO materials are extensively researched for their application in optoelectronic and nanoelectronic devices [4], photocatalysis [5], piezoelectric devices [6], sensors [7], surface acoustic wave devices, as antireflection coatings and windows in solar cells [8], and as transparent electrodes [9].

Doping ZnO is a successful method for modifying and improving its electronic, chemical, optical, and morphological properties, such as modifying grains' sizes and shapes, porosity, smoothness, etc. For example, doping ZnO with trivalent donor dopants such as Al^{3+}, Ga^{3+}, and In^{3+} on Zn^{2+} sites typically results in improved n-type conductivity [10], meanwhile, p-type ZnO can be achieved by introducing nitrogen (N), phosphorus (P), arsenic (As), antimony (Sb), and lithium (Li) dopants [11]. The ZnO band gap can be tailored by doping with such materials as Mg [12]. Rare-earth metal doping of ZnO is reported to be an effective way for the adjustment and control of gas-sensing efficiency [13]. Generally, doping alters the ZnO-based materials in view of different applications by enhancing the desired properties through choosing the appropriate dopant.

Lithium-doped ZnO structures are interesting for scientific research, as Li doping can induce enhanced optical, electrical, magnetic, and photoluminescence properties [14]. ZnO:Li has been reported to exhibit improved crystal quality, high sensitivity as a UV sensor [15], piezoelectric response [16], p-type conductivity [17], etc. The evident photocatalytic properties of ZnO:Li have also been reported [18].

Undoped ZnO films exhibit n-type conductivity as a consequence of deviation from their stoichiometry and of the presence of native donor defects, such as zinc interstitial (Zn_i) and oxygen vacancy (V_O) [19]. For different applications, such as p–n junction-based devices or bipolar device applications, it is necessary to obtain stable p-type oxide semiconductors [19,20]. There are two ways of inducing p-type conductivity in ZnO: one is doping with group I elements (Li, Na, K) at the Zn site; the other is substitution at the O site by group V elements (N, P, Sb) [19]. Lithium is a prospective dopant candidate, as it possesses a small ionic radius (0.68 Å) that is very close to the ionic radius of Zn (0.74 Å) [21,22]. Experimentally, it has been shown that Li doping provokes p-type semiconductivity by creating deep acceptor levels due to the Li atoms occupying Zn sites in the wurtzite host lattice and the Li^{1+} ions acting as shallow acceptors [14]. On the other hand, Li^{1+} ions can also occupy the interstitial positions (Li_i), where they become electron donors [23,24].

ZnO:Li films have been fabricated by numerous deposition techniques, including pulsed laser deposition [25], spray pyrolysis [26], sol–gel [15], electron beam evaporation [27], magnetron sputtering [28], etc. The sol–gel method exhibits several advantages, such as control of the film composition, easy film fabrication on large-area substrates, and low cost [29,30].

In this work, we present the sol–gel deposition of Li-doped ZnO thin films. ZnO:Li films were obtained by the spin-coating method on quartz and silicon substrates. The prepared sol solutions (Zn and mixed Zn/Li sols) remained stable for a period of five months, which is a very good technological achievement. In addition, a detailed study of the properties covering a wide temperature range, from 300 to 600 °C, is performed. The obtained samples are structurally characterized using XRD analysis and FTIR spectroscopy. Their optical properties are analyzed using UV-VIS spectroscopy in the spectral range of 240–1800 nm. The change in the optical band gap and refractive index with annealing is discussed. The influence of a lithium dopant on the ZnO film morphology is revealed.

2. Materials and Methods

A sol–gel spin-coating approach was applied to preparing ZnO and ZnO:Li thin films. A 0.4 M Zn sol solution was prepared: First, zinc acetate dihydrate $Zn(CH_3COO)_2{\cdot}2H_2O$ (Riedel de Haen, Hannover, Germany) was dissolved in absolute ethanol (Merck KgaA Darmstadt, Germany, absolute for analysis) [31]. Secondly, monoethanolamine (MEA, Fluka AG, Buchs, Switzerland, 98%) was added to bring the MEA/Zn molar ratio to unity. Finally, lithium nitrate $LiNO_3$ (Sigma-Aldrich Chemie GmbH, Taufkirchen, Germany) was added to reach a 0.04:1 Li:Zn molar ratio. Then, the solution was stirred using a magnetic stirrer (MS-H280-PRO, DLAB Scientific Co., Ltd., Beijing, China) at 55 °C for 1 h, followed by treatment in an ultrasonic cleaner (ELMA Elmasonic Easy 10H, Elma Schmidbauer GmbH, Singen, Germany) at 45 °C for 2 h. The obtained sols (pure and mixed solutions) were transparent with no precipitations. They were stable for a period of 5 months.

ZnO and ZnO:Li films were deposited on cleaned substrates by spin-coating (spin coater P 6708, PI-KEM Limited, Staffordshire, UK) at a 4000-rpm rotational speed for 30 s. The substrate cleaning included the following steps: cleaning in acetone, then treatment in ethanol (both steps were performed in an ultrasonic bath at 45 and 60 °C, respectively), and rinsing in double-distilled water. The final films were obtained after repeating the spin-coating process five times and, after each cycle of spin-coating, the substrates were preheated at a temperature of 300 °C for 10 min in a chamber furnace (chamber furnace, Tokmet—TK Ltd., Varna, Bulgaria) to evaporate the solvent and to decompose and remove the organic compounds. The sol–gel ZnO and ZnO:Li films were annealed at 300, 400, 500 and 600 °C for 1 h in air. Si wafers (FZ, p-type, resistivity 4.5–7.5 Ω, orientation <100>) were used for XRD, FESEM and FTIR studies, and UV-graded quartz-glass substrates (thickness 1 mm ± 0.1), for optical measurements.

The films' thickness was measured using an LEF 3 M laser ellipsometer (Siberian Branch of the Russian Academy of Sciences, Novosibirsk) with a He-Ne laser at the wavelength of 632.8 nm. The films' thickness values were 160 and 165 nm for ZnO and ZnO:Li films, respectively. These values were close so that the comparison of the films' optical and structural properties was appropriate.

The FTIR spectra were taken by an IR Prestige-21 FTIR spectrophotometer (Shimadzu Corporation, Kyoto, Japan) in the spectral range 350–4000 cm^{-1} (resolution of 4 cm^{-1}) using a bare Si wafer as background. The X-ray diffraction patterns were recorded by a Bruker D8 XRD diffractometer (Bruker AXS GmbH, Karlsruhe, Germany) using a Cu anode (Kα radiation) at a grazing angle of 1°, a step time of 2 s and a step of 0.04°. The optical spectra were recorded using a Shimadzu 3600 UV–VIS–NIR double-beam spectrophotometer (Shimadzu Corporation, Kyoto, Japan) in the 240–1800 nm spectral range and at a resolution of 0.1 nm. The transmittance was measured against air. The reflectance spectra were taken by using the specular reflectance accessory (at a 5° incidence angle) with an Al-coated mirror as reference. A four-point probe (model FPP-100, Veeco Instruments Inc.) was used for determining the samples' sheet resistance.

The films' morphologies were studied by field emission scanning electron (FESEM) microscopy (Philips XL 30FEG-ESEM, FEI, FEI Europe B.V., Zaventem, Belgium). A Au coating was deposited over the samples' surfaces before the measurements.

3. Results and Discussions

3.1. FTIR Investigation

FTIR spectroscopy is a technique used for identification of chemical bonds and functional groups [14]. The shapes and intensities of the absorption features depend on the sample's crystallinity, chemical composition, impurities, stress, and morphology, as well as on the crystallite's size and shape [32]. Figure 1a presents FTIR spectra of ZnO:Li films annealed at different temperatures; Figure 1b shows a comparison of the spectra of a ZnO and a ZnO:Li film after annealing at the highest temperature (600 °C). Introducing a dopant into the ZnO lattice can change the characteristic IR lines and give rise to new bands.

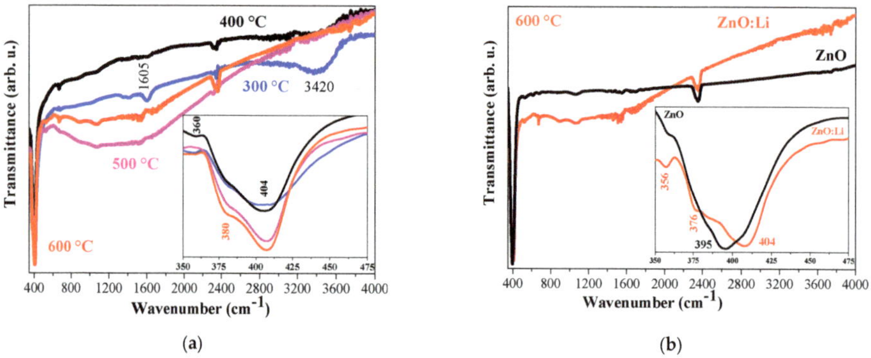

Figure 1. FTIR spectra of (**a**) ZnO:Li films treated at 300, 400, 500 and 600 °C and (**b**) comparison of the FTIR spectra of ZnO and ZnO:Li films annealed at 600 °C. The insets present the enlarged spectral region 350–475 cm^{-1}, where the main absorption bands appear.

The ZnO:Li film treated at 300 °C showed a strong and broad band at 3420 cm^{-1}, which was assigned to the stretching modes of hydroxyl groups. The corresponding OH bending vibrations were manifested by the characteristic line at 1605 cm^{-1} [14].

The hydroxyl groups' absorption bands vanished as the annealing temperature was increased, and the two IR lines at 3420 and 1605 cm^{-1} disappeared after the 500 °C thermal treatment. The absorption lines at 2346 and 2378 cm^{-1} observed in the spectra were related to CO_2 since the FTIR spectra were recorded in air [33]. The hydroxyl absorption bands were absent in the undoped ZnO spectra, even for the sample treated at the lowest annealing temperature (300 °C).

The characteristic metal-oxide absorption bands arising from inter-atomic vibrations generally appeared in the fingerprint IR region, i.e., below 1000 cm^{-1} [34]. The main absorption band of ZnO:Li films was observed at 404 cm^{-1}; its intensity rose with the annealing temperature (as can be seen in the inset in Figure 1a). This band was associated with the stretching Zn–O vibrations [34,35].

The comparison between the undoped and doped samples (Figure 1b) revealed that Li doping shifted the strongest IR line from 395 to 404 cm^{-1}. The absorption band of the ZnO:Li sample had an asymmetrical shape and a clear feature at 376 cm^{-1}. Another peak at 356 cm^{-1} was also seen. These lines were attributed to Zn-O stretching vibrations [32]. For all of the annealing temperatures, the main band of the Li-containing ZnO films was wider, proving that the Li ions were embedded into the host lattice [33].

The FTIR study revealed that adding Li was manifested by changes in the absorption features. Raising the annealing temperatures led to stronger IR lines, suggesting that the film crystallinity was improved. XRD analysis was applied for revealing the crystallinity evolution with annealing.

3.2. XRD Structural Study

Figure 2 presents the XRD patterns of Li-doped ZnO films deposited on Si substrates; the XRD patterns of undoped ZnO and ZnO:Li films annealed at 600 °C are shown in Figure 3. The X-ray diffraction patterns matched well with the standard values of hexagonal wurtzite ZnO (JCPDS PDF card no. 00-036-1451) and confirmed the sol–gel films' crystallization. The structural results revealed the polycrystalline nature of all samples.

All peaks observed could be indexed to zinc oxide. No diffraction peaks related to Li-containing phases were detected; this may suggest the incorporation of lithium in the ZnO crystal structure, but the low Li concentration could also be responsible for the absence of a signature in the XRD patterns.

Table 1 summarizes the ZnO lattice parameters, the crystallite size, the dislocation density and the c/a ratio of undoped ZnO and ZnO:Li films. These parameters were determined using the equations given in [36,37]. The crystallites' average size (d) was estimated by the Scherrer equation [37] using the full width at half maximum (FWHM) of the (100), (002) and (101) diffraction peaks. In order to allow a comparison with reference data, we also quoted $1/d^2$, which many authors call the dislocation density [36], although this formula is a simplification of Williamson and Smallman's work on annealed and cold-worked metals [38].

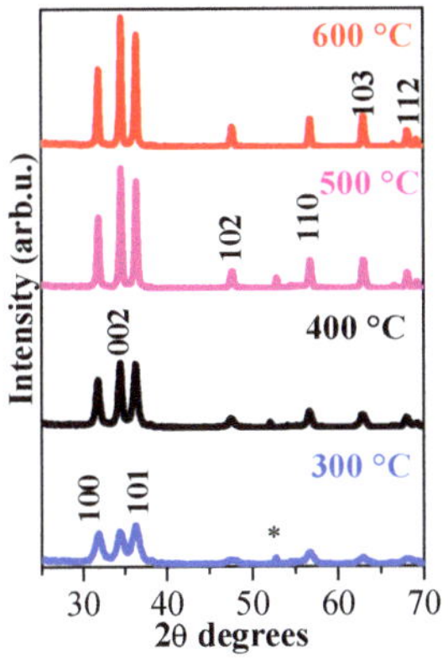

Figure 2. XRD patterns of sol–gel ZnO:Li films treated at 300–600 °C. The asterisk (*) marks a peak arising from the Si substrate.

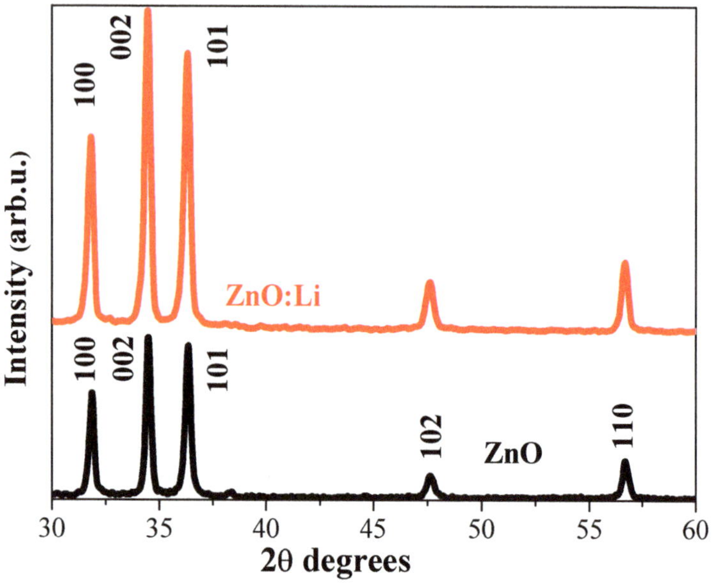

Figure 3. XRD patterns of sol-gel ZnO and ZnO:Li films treated at 600 °C.

Table 1. Crystallite size (d), dislocation density ($1/d^2$), lattice constant (a), (c) and c/a ratio of undoped ZnO and ZnO:Li films estimated using XRD data.

$T_{annealing}$ [°C]	Parameter	Undoped ZnO	ZnO:Li
300	d [nm]	11.8 (7)	11.6 (7)
	$1/d^2 \times 10^{-4}$ [1/nm^2]	71	74
	a [Å]	3.241 (3)	3.243 (3)
	c [Å]	5.199 (6)	5.201 (6)
	c/a ratio	1.604	1.604
400	d [nm]	15 (1)	19 (1)
	$1/d^2 \times 10^{-4}$ [1/nm^2]	44	28
	a [Å]	3.243 (3)	3.244 (3)
	c [Å]	5.199 (6)	5.200 (6)
	c/a ratio	1.603	1.603
500	d [nm]	30 (1)	29 (1)
	$1/d^2 \times 10^{-4}$ [1/nm^2]	11	12
	a [Å]	3.244 (3)	3.243 (3)
	c [Å]	5.197 (6)	5.197 (6)
	c/a ratio	1.602	1.603
600	d [nm]	36 (2)	31 (1)
	$1/d^2 \times 10^{-4}$ [1/nm^2]	8	10
	a [Å]	3.243 (3)	3.243 (3)
	c [Å]	5.193 (6)	5.194 (6)
	c/a ratio	1.601	1.602

The uncertainties in the cell parameters and the crystallite size are estimated by assuming a 0.02° uncertainty in the peak positions and in the FWHM.

Increasing the annealing temperature induced an increase in the crystallite size in both cases. The differences in the crystallite size between undoped and Li-doped ZnO were too small to draw conclusions, although the fact that the crystallite size was somewhat larger for undoped ZnO at 600 °C agrees with the trend observed by Fujihara et al. [30] for the sol–gel films.

As noted above, both undoped ZnO and ZnO:Li films crystallized in a wurtzite phase. The wurtzite structure had a hexagonal unit cell with two lattice parameters, *a* and *c*, in the ratio of c/a = 1.633 (an ideal wurtzite crystal) and belonged to the $P6_3mc$ space group. In a real ZnO crystal, the wurtzite structure deviates from the ideal arrangement—the *c/a* ratio is close to 1.60. Many factors have been reported to affect the lattice parameters and the c/a ratio: (i) lattice distortion, (ii) impurities and defects (iii) differences in the ionic radii of O^{2-}, Zn^{2+}, and Li^{+}, (iv) external strains induced by the substrate and the temperature, (v) electrostatic interactions between the ions in the lattice (these interactions influence the optimal distances between the ions in undoped and doped zinc oxide) [39], and (vi) defects in the real lattice [40]. However, in the present case, the lattice parameter variations caused by Li doping and thermal treatment were not significant given the uncertainty in the peak positions. Indeed, the percentage of lithium was rather small. Hjiri et al. [41] also did not observe any (002) peak shifts for Li doping concentrations up to 3 at%. Both Song et al. [16] and Jeong et al. [28] reported a small decrease in the *c* parameter in the case of a Li-doped film prepared by magnetron sputtering, a technique allowing one to reach higher levels of lithium trapped in the ZnO matrix in comparison with the sol–gel routes. The c/a ratio values of sol–gel ZnO and ZnO:Li films differed slightly from the reference value of 1.602 of ZnO (PDF card no. 00-036-1451). A more significant effect of Li addition was observed regarding the intensity distribution between the (100), (002) and (101) peaks illustrated in Figure 4 as the texture coefficient (*TC*) calculated from the following equation [42]:

$$TC\ (hkl) = \frac{I_{(hkl)}/I_{o(hkl)}}{N^{-1}\sum_N I_{(hkl)}/I_{o(hkl)}} \quad (1)$$

where *I(hkl)* is the measured relative intensity; $I_o(hkl)$ is the (*hkl*) plane standard intensity; and *N* is the number of diffraction lines. The standard intensities for (100), (002) and (101) were taken from PDFS card 01-070-8070. Considering the small 2θ range for these three reflections, combining the experimental intensities collected in grazing incidence with the standard intensities in a Bragg–Brentano geometry did not affect the qualitative conclusions.

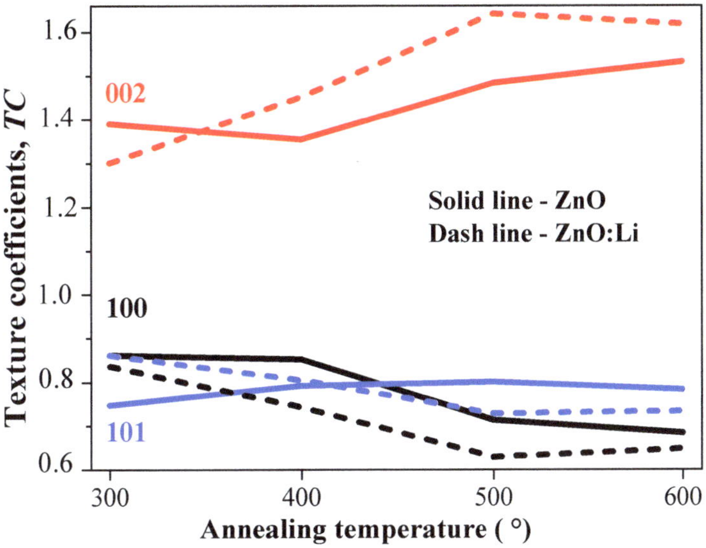

Figure 4. Comparison of the texture coefficients of (100), (002), (101) peaks of ZnO and ZnO:Li films as a function of the annealing temperature.

Figure 4 presents the annealing temperature effect on the estimated texture coefficients (*TC*) of the (100), (002), (101) diffraction planes. It is known that the *TC* value represents the texture of the particular plane—its exceeding unity implies a preferred crystallite growth with this orientation. The ZnO and ZnO:Li films considered here both had *TC* (101) and *TC* (100) below unity, which indicates that the preferred orientation was along another direction. The *TC* (100) values were near 0.8 for ZnO and in the range 0.73–0.86 for ZnO:Li (Figure 4). *TC* (002) was greater than unity for both undoped (1.36–1.53) and Li-doped ZnO (1.30–1.64). Thus, it was found that Li doping brings about a higher degree of orientation along the 002 plane for annealing temperatures in the 400–600 °C range. The highest value of 1.64 was measured for the ZnO:Li film treated at 500 °C.

The XRD studies showed that Li doping had a slight impact on the wurtzite structure of the ZnO films. The sol–gel ZnO:Li films crystallized with a preferential growth orientation along the *c*-axis. The hexagonal wurtzite structure was maintained without the formation of an additional phase. The small Li doping level did not modify significantly the lattice parameters and the crystallite size, while the thermal treatment favored the growth of larger crystallites.

3.3. Film Surface Morphology

The ZnO:Li surface morphology evolution with the annealing was studied by FESEM. The samples were deposited on Si wafers and treated at 300 and 600 °C. Figure 5 shows the FESEM images (at two magnifications) of the sol–gel Li-doped ZnO film annealed at 300 °C. The surface morphology has a wrinkle-type surface structure with thinner and thicker wrinkles (at the magnification of 20,000, Figure 5a). The formation of wrinkles on the surface of sol–gel ZnO coatings was reported previously in [43,44]. Some authors [45] proposed that voids appear as a result of elimination of residual organic solvents during the preheating and annealing procedures, so that the stress imbalance arising in the films causes wrinkle-like surface features [46].

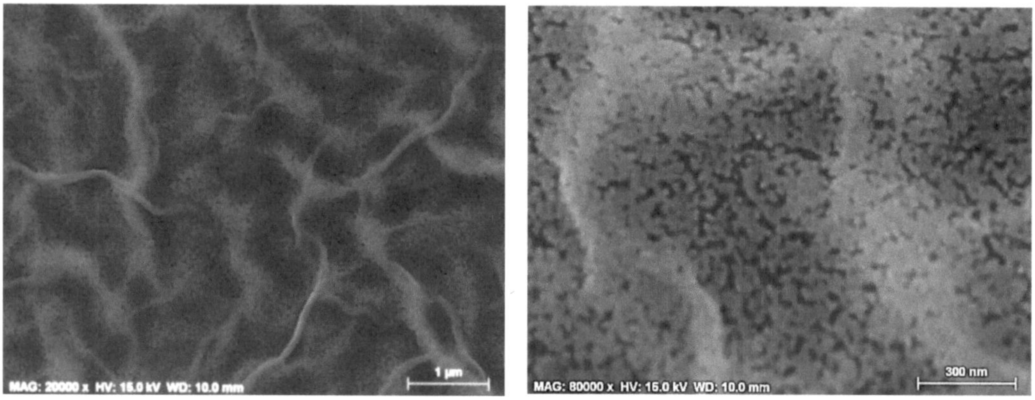

(a) (b)

Figure 5. FESEM micrographs of a ZnO:Li film deposited on Si and treated at 300 °C. The images show the film surface at (**a**) 20,000 and (**b**) 80,000 magnification.

The image at higher magnification (Figure 5b) illustrates a rather porous structure with tiny grains with sizes below 20 nm that are difficult to distinguish. The average crystallites size determined by XRD was 11 nm, as estimated from the diffraction planes (112), (103), (110), (101), (002) and (100). The FESEM study revealed similar or slightly bigger grains; however, as known, the crystallite size is assumed to be the size of a coherently diffracting domain and does not represent exactly the particle size [40].

The annealing at 600 °C produced different morphologies. A comparison of the FESEM images of undoped ZnO and ZnO:Li films obtained under the same technological conditions is given in Figure 6. The lithium dopant provoked a significant change in the film morphology. The undoped ZnO film exhibits wrinkles with a very porous structure, with the grains having irregular shapes in sizes varying from 30 to 100 nm (Figure 6a,b). Annealing the sol–gel ZnO:Li film at the highest temperature resulted in a wrinkle-type surface structure (Figure 6c) with well-defined fibers (or wrinkles) consisting of distinct nanoparticles (Figure 6c). The grains of the ZnO:Li thin film annealed at 600 °C clung closely to each other. The highest annealing temperature transformed the porous film morphology to a denser structure without pores and with well-defined grains, some of which were of spheroidal shape. The grain size (as determined from the micrograph with 80,000 magnification, Figure 6d) varied from 30 nm to 110 nm.

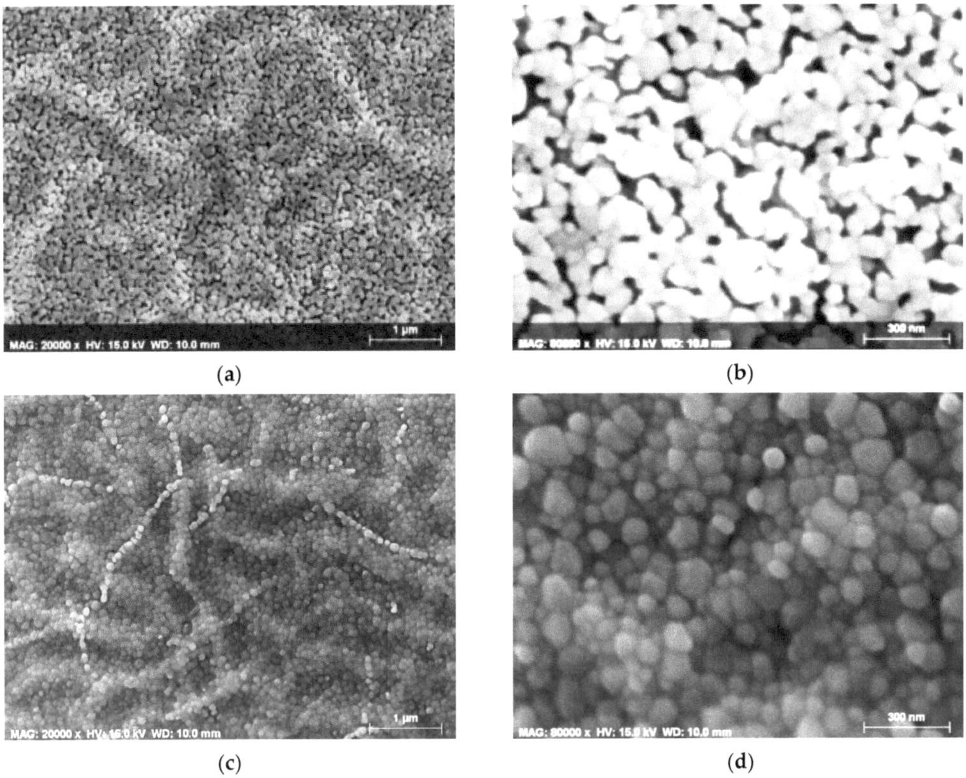

Figure 6. FESEM micrographs of undoped ZnO and ZnO:Li films deposited on Si and treated at 600 °C. The images show the film surface morphology of (**a**) ZnO film at 20,000; (**b**) ZnO film at 80,000 magnification; (**c**) ZnO:Li film at 20,000; and (**d**) ZnO:Li film at 80,000 magnification.

The morphology of undoped ZnO is porous, while the ZnO:Li films manifest a denser structure with closely packed grains. The FESEM images confirmed that the ZnO:Li films are nanostructured materials, with the thermal treatment causing structural and surface modifications of the films.

3.4. Optical Characterization

Figure 7 presents the spectra of ZnO:Li thin films annealed from 300 to 600 °C and recorded in the 200–1800 nm spectral range. The thermal treatment reduced slightly the films' transparency. The difference appeared after the annealing at 600 °C. The ZnO:Li films annealed at temperatures above 400 °C showed exciton absorption peaks around 340 nm, confirming the good crystallinity of the films [46]. Figure 8a provides the average values of the transmittance and reflectance in the visible spectral region (450–750 nm) of undoped and Li-doped ZnO films. Following the post-annealing, the average optical transmittance slightly decreased. The average reflectance was below 9% and changed weakly with the annealing temperature. The average transmittance and reflectance of the bare quartz substrate were 93.5% and 6.7%, respectively. The Li doping improved the optical transparency. The ZnO:Li films' reflectance was lower than that of the ZnO films. The undoped films exhibited a trend of increasing the reflectance at higher annealing temperatures. The FESEM analysis and the microscopic images revealed denser and smoother surfaces of the sol–gel doped films. The smooth morphology resulted in an improved average transmittance of the Li-doped ZnO films.

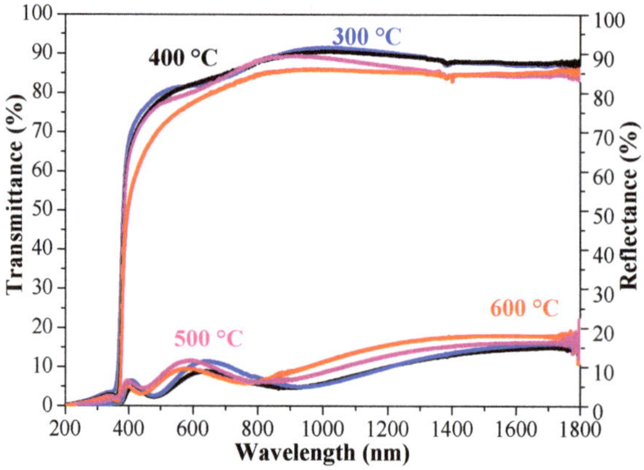

Figure 7. Transmittance and reflectance spectra of ZnO:Li films annealed at temperatures of 300–600 °C. The substrate used is quartz.

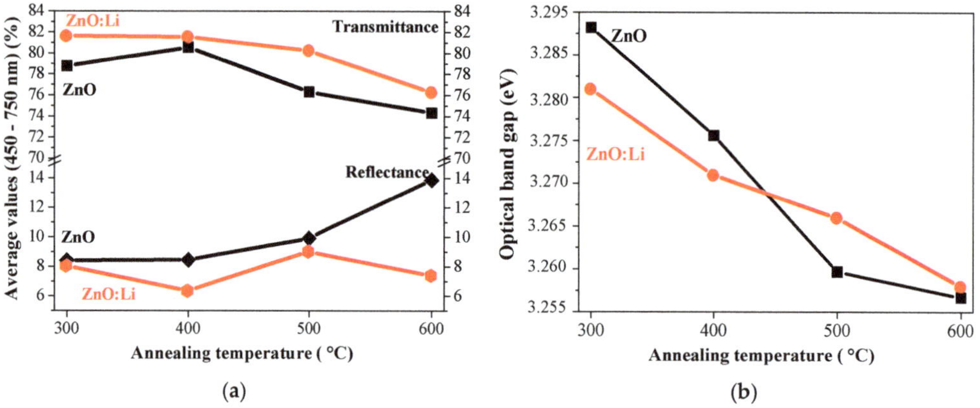

Figure 8. Comparison of (**a**) the average values of transmittance and reflectance in the visible spectral range 450–750 nm; and (**b**) the optical band gaps of ZnO and ZnO:Li films as a function of the annealing temperature.

The optical band gap is derived from the first derivative of the transmittance versus the energy [47]. The estimated optical band gap (E_g) values for ZnO and ZnO:Li films are shown in Figure 8b as a function of the annealing temperatures; the results are in good agreement with the literature data for ZnO-based materials [48,49]. A narrowing trend was seen in the E_g of the ZnO and ZnO:Li films annealed at the higher temperatures. The optical band gap can be influenced by several factors, including crystallization, grain sizes, structural parameters, impurities, etc. [50]. The decrease in the band gap energy with annealing correlated with the growth of crystallites of greater size [50].

The lower values of the optical band gap of ZnO:Li films compared to undoped ZnO after annealing at 300 and 400 °C can be related to the change in the lattice parameters and the formation of a tail band [51]. The higher-temperature thermal treatments (at 500 and 600 °C) reversed this tendency, as ZnO:Li samples have wider optical band gaps than those of ZnO (Figure 8b). The widening of E_g can be due to the Burstein–Moss effect and to Li occupying interstitial sites in ZnO [51].

The refractive index, n, is an important parameter characterizing the optical properties of a thin film. The refractive index for the wavelengths ranging from 240 to 1700 nm was calculated using the measured reflectance spectra of the ZnO and ZnO:Li films. The following relation was used [52]:

$$R = \frac{(n-1)^2 + k^2}{(n+1)^2 + k^2} \quad (2)$$

where k is the extinction coefficient and R is the reflectance. If $k << n$, then:

$$n = \frac{1+\sqrt{R}}{1-\sqrt{R}} \quad (3)$$

Figure 9 displays the obtained values for the refractive index of undoped ZnO (Figure 9a) and ZnO:Li (Figure 9b) films annealed at 300–600 °C.

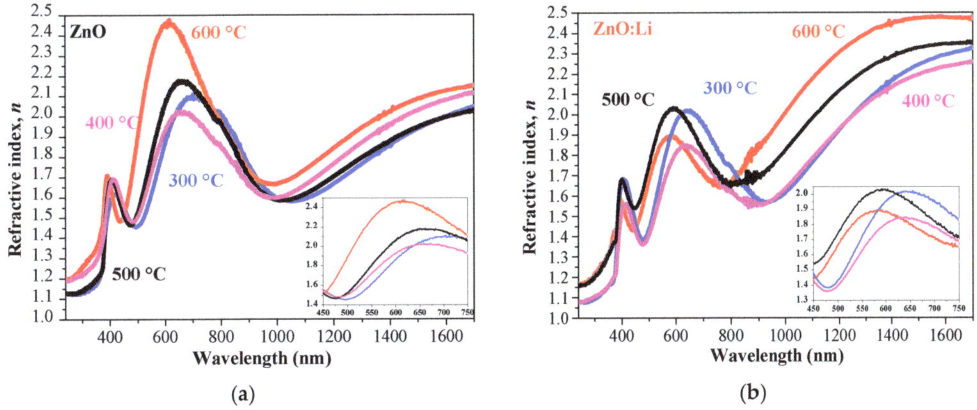

Figure 9. Refractive index, n, values for (**a**) undoped ZnO and (**b**) ZnO:Li films after annealing at 300–600 °C. The inset figures show the refractive index values in the spectral range 450–750 nm.

The values of n depend strongly on the wavelength. The wavelengths of the undoped ZnO films revealed that n increases as the annealing temperature is increased, especially in the visible spectral range (Figure 9a). A similar tendency was observed for the ZnO:Li films in the near IR region above 1000 nm. For wavelengths below 750 nm, the refractive index of the doped samples treated at 300 and 500 °C reached its highest values. Generally, the refractive index of the ZnO:Li is lower than that of the ZnO films in the wavelength region 240–990 nm. In the NIR spectral range, the ZnO:Li films' n values exceeded those of ZnO. The decrease in the refractive index with Li doping in the range 300–900 nm has been reported by other authors [53]. In contrast, other researchers [50,54] reported higher refractive index values for Li-doped ZnO films. The refractive index of single-crystal ZnO is 2.047–2.063 at 500 nm [55]. The obtained n values are in agreement with the reported values. It must be noted that there is a large dispersion in the reported values of the refractive index of ZnO-based materials depending on the deposition methods, the technological conditions, stoichiometry, crystallinity, dopants and packing density [55].

The optical characterization of the studied ZnO:Li films demonstrated an improved transparency (up to 82%) and a change in the optical band gap (a narrowing from 3.281 to 3.258 eV with the temperature). The refractive index reduction in the visible spectral range was caused by incorporation of Li in the zinc oxide host lattice.

3.5. Electrical Properties

The transparent conductive films' performance can be evaluated by using the concept of the figure of merit (*FOM*) (Haacke [56]) and Equation (4):

$$FOM = \frac{{T_{average}}^{10}}{R_{sheet}} \tag{4}$$

where $T_{average}$ is the average transmittance in a certain spectral region (in our case, 450–750 nm) and R_{sheet} is the sheet resistance of the samples. The FOM yields a numerical value derived from the two most important parameters of a transparent conductor, namely, transmittance and sheet resistance. It is known that these parameters in semiconductors are interdependent—this is why the figure of merit is used [57]. Improving the optical transparency (or electrical conduction) will result in a reduction in the electrical conduction (or transmittance) [57].

Table 2 presents the FOM values estimated for the sol–gel Li-doped ZnO films treated at 300, 500 and 600 °C. The values quoted for the average transmittance were evaluated in the visible spectral range 450–750 nm by extracting the quartz substrate transmittance. The sheet resistance was measured using the four-probe technique.

Table 2. Average transmittance, $T_{average}$, (estimated for the spectral range 450–750 nm), sheet resistance, R_{sheet}, and figure of merit (*FOM*) of ZnO:Li films treated at different temperatures.

$T_{\text{annealing}}$ [°C]	$T_{average}$ [%]	R_{sheet} [Ω/sq]	$FOM \times 10^{-4}$ [Ω^{-1}]
300	88.10	380	7.42
500	86.74	396	6.08
600	82.75	250	6.02

In order to highlight the effect of the sol–gel method on the *FOM* of the ZnO thin films, a comparison of the properties of thin films with similar doping in terms of dopant and doping level is necessary. Unfortunately, to the best of our knowledge, data on Li-doped ZnO thin films are not available. Table 3 presents a comparison of reported sheet resistance and *FOM* values for ZnO-based films prepared by different deposition methods [43,58–62]. As can be seen, the R_{sheet} and *FOM* values vary widely. Comparing the results of the sol–gel ZnO:Li films studied in the present work, it is seen that they approached the reported values, but there are doped and undoped ZnO films with better electrical properties. It must be emphasized that this study was focused on the effects of Li doping and the annealing temperatures on the structural, optical and morphological properties of ZnO films.

Table 3. Comparison of $T_{average}$, R_{sheet} and figures of merit (*FOM*) of undoped and doped ZnO films, prepared by different deposition methods.

Material	Deposition Method	$T_{average}$ [%]	R_{sheet} [Ω/sq]	*FOM* [Ω^{-1}]	Reference
ZnO	Spray pyrolysis	96.30	388	1.76×10^{-3}	[58]
ZnO:Al	Sol–gel	84.19		0.94×10^{-4}	[43]
ZnO:Al:In	RF sputtering	88.00 (550 nm)	9.6	2.65×10^{-2}	[59]
ZnO:Al (implanted) ZnO	Sol–gel	82.20	156	9.03×10^{-4}	[60]
ZnO	Sol–gel	91.75	1950	2.17×10^{-4}	[61]
ZnO:Ga	Atmospheric pressure plasma jet	83.40 (550 nm)	12	1.48×10^{-2}	[62]
ZnO:Li	Sol–gel	82.75	250	6.02×10^{-4}	This work

Regarding the electrical properties of sol–gel ZnO:Li films, further research is needed on optimizing the preheating and post-annealing procedures and the films' thickness and transparency. In this respect, the results obtained in the present work are promising.

4. Conclusions

Thin films of lithium-doped zinc oxide were successfully prepared on Si and quartz substrates using the simple sol–gel spin-coating method. The FTIR analysis showed that including Li in ZnO thin films markedly modified and shifted the absorption bands. No absorption bands related to Li oxides were detected. The XRD patterns demonstrated the polycrystalline structure of the ZnO:Li films, which kept their wurtzite structure; no Li-containing crystalline phases were detected. The crystallite sizes increased as the annealing temperature was increased, and doping with lithium changed the temperature evolution of the texture coefficients (*TC*) in ZnO. The preferential (002) orientation was enhanced in the ZnO:Li films compared to the undoped samples. Introducing the dopant annealing modified the surface morphology. The FESEM observations showed a more uniform and compact structure of the ZnO:Li films, which were composed of closely packed spherical nanograins, in contrast to the very porous structure of the undoped ZnO consisting of randomly distributed irregular grains. The Li dopant improved the ZnO optical transparency across the visible spectral range, as the average transmittance was found to be higher than 80% for the ZnO:Li films annealed at 300–500 °C. The preliminary electrical study yielded encouraging results. In summary, we can conclude that lithium doping affects the structural, optical and morphological properties of sol–gel prepared ZnO thin films. Further, the ZnO:Li nanostructured thin films deposited using the cost-effective sol–gel process are promising candidates for practical technological applications.

Author Contributions: Conceptualization, T.I. and A.H.; methodology, T.I., A.H., T.K. and B.V.; validation, T.I. and R.C.; formal analysis, T.I., T.K. and A.H.; investigation, T.I., A.H., T.K., B.V. and R.C.; data curation, T.I., T.K. and B.V.; writing—original draft preparation, T.I., A.H., T.K. and B.V.; writing—review and editing, T.I.; visualization, T.K. and R.C.; supervision, T.I.; project administration, T.I. All authors have read and agreed to the published version of the manuscript.

Funding: This research received no external funding.

Data Availability Statement: Data are contained within the article.

Conflicts of Interest: The authors declare no conflict of interest.

References

1. Morales, C.; Leinen, D.; del Campo, A.; Ares, J.R.; Sanchez, C.; Flege, J.I.; Gutierrez, A.; Prieto, P.; Soriano, L. Growth and characterization of ZnO thin films at low temperatures: From room temperature to −120 °C. *J. Alloys Comp.* **2021**, *884*, 161056. [CrossRef]
2. Wang, Z.; Luo, C.; Anwand, W.; Wagner, A.; Butterling, M.; Rahman, M.A.; Phillips, M.R.; Ton-That, C.; Younas, M.; Su, S.; et al. Vacancy cluster in ZnO films grown by pulsed laser deposition. *Sci. Rep.* **2019**, *9*, 3534. [CrossRef] [PubMed]
3. Borysiewicz, M.A. ZnO as a Functional Material, a Review. *Crystals* **2019**, *9*, 505. [CrossRef]
4. Muchuweni, E.; Sathiaraj, T.S.; Nyakotyo, H. Synthesis and characterization of zinc oxide thin films for optoelectronic applications. *Heliyon* **2017**, *3*, e00285. [CrossRef] [PubMed]
5. Mohamed, K.M.; Benitto, J.J.; Vijaya, J.J.; Bououdina, M. Recent Advances in ZnO-Based Nanostructures for the Photocatalytic Degradation of Hazardous, Non-Biodegradable Medicines. *Crystals* **2023**, *13*, 329. [CrossRef]
6. Pandey, K.; Dutta, J.; Brahma, S.; Rao, B.; Liu, C.P. Review on ZnO-based piezotronics and piezoelectric nanogenerators: Aspects of piezopotential and screening effect. *J. Phys. Mater.* **2021**, *4*, 044011. [CrossRef]
7. Khatibani, A.B. Characterization and Ethanol Sensing Performance of Sol-Gel Derived Pure and Doped Zinc Oxide Thin Films. *J. Electron. Mater.* **2019**, *48*, 3784–3793. [CrossRef]
8. Wibowo, A.; Marsudi, M.A.; Amal, M.I.; Ananda, M.B.; Stephanie, R.; Ardy, H.; Diguna, L.J. ZnO nanostructured materials for emerging solar cell applications. *RSC Adv.* **2010**, *10*, 42838–42859. [CrossRef]
9. Hála, M.; Fujii, S.; Redinger, A.; Inoue, Y.; Rey, G.; Thevenin, M.; Deprédurand, V.; Weiss, T.P.; Bertram, T.; Siebentritt, S. Highly conductive ZnO films with high near infrared transparency. *Prog. Photovolt. Res. Appl.* **2015**, *23*, 1630–1641. [CrossRef]
10. Zhao, D.; Li, J.; Sathasivam, S.; Carmalt, C.J. n-Type conducting P doped ZnO thin films via chemical vapor deposition. *RSC Adv.* **2020**, *10*, 34527–34533. [CrossRef]

11. Pathak, T.K.; Kumar, V.; Swart, H.C.; Purohit, L.P. P-type conductivity in doped and codoped ZnO thin films synthesized by RF magnetron sputtering. *J. Modern Opt.* **2015**, *62*, 1368–1373. [CrossRef]
12. Mia, M.N.H.; Mia, M.F.; Pervez, M.F.; Khalid Hossain, M.K.; Rahman, M.R.; Uddin, M.J.; Al Mashud, M.A.; Ghosh, H.K.; Hoq, M Influence of Mg content on tailoring optical bandgap of Mg-doped ZnO thin film prepared by sol-gel method. *Results Phys.* **2017** *7*, 2683–2691. [CrossRef]
13. Sayago, I.; Santos, J.P.; Sánchez-Vicente, C. The Effect of Rare Earths on the Response of Photo UV-Activate ZnO Gas Sensors *Sensors* **2022**, *22*, 8150. [CrossRef] [PubMed]
14. Punia, K.; Lal, G.; Dolia, S.N.; Kumar, S. Defects and oxygen vacancies tailored structural, optical, photoluminescence and magnetic properties of Li doped ZnO nanohexagons. *Ceram. Int.* **2020**, *46*, 12296–12317. [CrossRef]
15. Lee, W.; Leem, J.-Y. Ultraviolet Photoresponse Properties of Li-Doped ZnO Thin Films Prepared by Sol–Gel Spin-Coating Method *J. Nanosci. Nanotechnol.* **2017**, *17*, 5697–5700. [CrossRef]
16. Song, M.; Liu, Y.; Yu, A.; Zhang, Y.; Zhai, J.; Wang, Z.L. Flexible Li-doped ZnO piezotronic transistor array for in-plane strain mapping. *Nano Energy* **2019**, *55*, 341–347. [CrossRef]
17. Khosravi, P.; Ebrahimi, S.A.S. Structural, Electrical and Optical Characterization of ZnO:Li Thin Films Prepared by Sol-Gel Spin Coating. *J. Ultrafine Grained Nanostruct Mater.* **2023**, *56*, 108–120.
18. Salah, M.; Azizi, S.; Boukhachem, A.; Khaldi, C.; Amlouk, M.; Lamloumi, J. Effects of lithium doping on: Microstructure morphology, nanomechanical properties and corrosion behaviour of ZnO thin films grown by spray pyrolysis technique. *J. Mater Sci. Mater. Elect.* **2019**, *30*, 1767–1785. [CrossRef]
19. Tsay, C.-Y.; Chiu, W.-Y. Enhanced Electrical Properties and Stability of P-Type Conduction in ZnO Transparent Semiconductor Thin Films by Co-Doping Ga and N. *Coatings* **2020**, *10*, 1069. [CrossRef]
20. Zagal-Padilla, C.K.; Gamboa, S.A. Role of native defects on the opto-electronic properties of p-type ZnO synthesized during the most straightforward method: Only water. *Appl. Phys. A* **2023**, *129*, 183. [CrossRef]
21. Scajev, P.; Durena, R.; Onufrijevs, P.; Miasojedovas, S.; Malinauskas, T.; Stanionyte, S.; Zarkov, A.; Zukuls, A.; Bite, I.; Smits, K Morphological and optical property study of Li doped ZnO produced by microwave-assisted solvothermal synthesis. *Mater. Sci Semicond. Process.* **2021**, *135*, 106069. [CrossRef]
22. Chirakkara, S.; Krupanidhi, S.B. Pulsed laser deposited ZnO/ZnO:Li multilayer for blue light emitting diodes. *J. Lumin.* **2011** *131*, 1649–1654. [CrossRef]
23. Jin, M.; Li, Z.; Huang, F.; Xia, Y.; Ji, X.; Wang, W. Critical conditions for the formation of p-type ZnO with Li doping. *RSC Adv* **2018**, *8*, 30868–30874. [CrossRef] [PubMed]
24. Ahmoum, H.; Boughrara, M.; Suait, M.S.; Kerouad, M. Effect of position and concentration of Li on ZnO physical properties Density functional investigation. *Chem. Phys. Lett.* **2019**, *719*, 45–53. [CrossRef]
25. Xiao, B.; Ye, Z.; Zhang, Y.; Zeng, Y.; Zhu, L.; Zhao, B. Fabrication of p-type Li-doped ZnO films by pulsed laser deposition. *Appl Surf. Sci.* **2006**, *253*, 895–897. [CrossRef]
26. Bornand, V.; Mezy, A. Morphological and ferroelectric studies of Li-doped ZnO thin films. *Mater. Lett.* **2013**, *107*, 357–360 [CrossRef]
27. Kafadaryan, E.A.; Petrosyan, S.I.; Hayrapetyan, A.G.; Hovsepyan, R.K.; Manukyan, A.L.; Vardanyan, E.S.; Zerrouk, A.F. Infrared 45° reflectometry of Li doped ZnO films. *J. Appl. Phys.* **2004**, *95*, 3005–3009. [CrossRef]
28. Jeong, S.H.; Park, B.N.; Lee, S.-B.; Boo, J.-H. Study on the doping effect of Li-doped ZnO film. *Thin Solid Films* **2008**, *516*, 5586–5589 [CrossRef]
29. Gartner, M.; Stroescu, H.; Mitrea, D.; Nicolescu, M. Various Applications of ZnO Thin Films Obtained by Chemical Routes in the Last Decade. *Molecules* **2023**, *28*, 4674. [CrossRef]
30. Fujihara, S.; Sasaki, C.; Kimura, T. Effects of Li and Mg doping on microstructure and properties of sol-gel ZnO thin films. *J. Eur Ceram. Soc.* **2001**, *21*, 2109–2112. [CrossRef]
31. Ivanova, T.; Harizanova, A.; Koutzarova, T.; Vetruyen, B. Study of ZnO sol–gel films: Effect of annealing. *Mater. Lett.* **2010**, *64* 1147–1149. [CrossRef]
32. Petrovic, Z.; Ristic, M.; Music, S. Development of ZnO microstructures produced by rapid hydrolysis of zinc acetylacetonate *Ceram. Int.* **2014**, *40*, 10953–10959. [CrossRef]
33. Mariammal, R.N.; Ramachandran, K. Increasing the Reactive Sites of ZnO Nanoparticles by Li Doping for Ethanol Sensing. *Mater Res. Express* **2019**, *6*, 015024. [CrossRef]
34. Istrate, A.-I.; Nastase, F.; Mihalache, I.; Comanescu, F.; Gavrila, R.; Tutunaru, O.; Müller, R. Synthesis and characterization of Ca doped ZnO thin films by sol–gel method. *J. Sol-Gel Sci. Technol.* **2019**, *92*, 585–597. [CrossRef]
35. Khan, M.F.; Ansari, A.H.; Hameedullah, M.; Ahmad, E.; Husain, F.M.; Zia, Q.; Baig, U.; Zaheer, M.R.; Alam, M.M.; Khan, A.M. et al. Sol-gel synthesis of thorn-like ZnO nanoparticles endorsing mechanical stirring effect and their antimicrobial activities Potential role as nano-antibiotics. *Sci. Rep.* **2016**, *6*, 27689. [CrossRef]
36. Bilgin, V. Preparation and characterization of ultrasonically sprayed zinc oxide thin films doped with lithium. *J. Electronic. Mater* **2009**, *38*, 1969–1978. [CrossRef]
37. Hussein, H.M. Photosensitive analysis of spin coated Cu doped ZnO thin film synthesized by hydrothermal method. *Results Opt* **2023**, *13*, 100543. [CrossRef]

38. Williamson, G.K.; Smallman, R.E., III. Dislocation densities in some annealed and cold-worked metals from measurements on the X-ray debye-scherrer spectrum. *Philos. Mag. J. Theor. Exper. Appl. Phys.* **1956**, *1*, 34–46. [CrossRef]
39. Wojnarowicz, J.; Chudoba, T.; Gierlotka, S.; Sobczak, K.; Lojkowski, W. Size Control of Cobalt-Doped ZnO Nanoparticles Obtained in Microwave Solvothermal Synthesis. *Crystals* **2018**, *8*, 179. [CrossRef]
40. Hsu, H.-P.; Lin, D.-Y.; Lu, C.-Y.; Ko, T.-S.; Chen, H.-Z. Effect of Lithium Doping on Microstructural and Optical Properties of ZnO Nanocrystalline Films Prepared by the Sol-Gel Method. *Crystals* **2018**, *8*, 228. [CrossRef]
41. Hjiri, M.; Aida, M.S.; Lemine, O.M.; El Mir, L. Study of defects in Li-doped ZnO thin films. *Mater. Sci. Semicond. Process.* **2019**, *89*, 149. [CrossRef]
42. Sharmila, B.; Singha, M.K.; Dwivedi, P. Impact of annealing on structural and optical properties of ZnO thin films. *Microeloect. J.* **2023**, *135*, 105759.
43. Khan, M.I.; Neha, T.R.; Billah, M.M. UV-irradiated sol-gel spin coated AZO thin films: Enhanced optoelectronic properties. *Heliyon* **2022**, *8*, e08743. [CrossRef] [PubMed]
44. Elsayed, I.A.; Afify, A.S. Controlling the Surface Morphology of ZnO Nano-Thin Film Using the Spin Coating Technique. *Materials* **2022**, *15*, 6178. [CrossRef] [PubMed]
45. Podia, M.; Tripathi, A.K. Structural, optical and luminescence properties of ZnO thin films: Role of hot electrons defining the luminescence mechanisms. *J. Lumin.* **2022**, *252*, 119331. [CrossRef]
46. Salam, S.; Islam, M.; Akram, A. Sol–gel synthesis of intrinsic and aluminum-doped zinc oxide thin films as transparent conducting oxides for thin film solar cell. *Thin Solid Films* **2013**, *529*, 242–247. [CrossRef]
47. Maache, A.; Chergui, A.; Djouadi, D.; Benhaoua, B.; Chelouche, A.; Boudissa, M. Effect of La doping on ZnO thin films physical properties: Correlation between strain and morphology. *Optik* **2019**, *180*, 1018–1026. [CrossRef]
48. Caglar, M.; Caglar, Y.; Aksoy, S.; Ilican, S. Temperature dependence of the optical band gap and electrical conductivity of sol-gel derived undoped and Li-doped ZnO films. *Appl. Surf. Sci.* **2010**, *256*, 4966–4971. [CrossRef]
49. Shohany, B.G.; Zak, A.K. Doped ZnO nanostructures with selected elements—Structural, morphology and optical properties: A review. *Ceram. Int.* **2020**, *46*, 5507–5520. [CrossRef]
50. Salah, M.; Azizi, S.; Boukhachem, A.; Khaldi, C.; Amlouk, M.; Lamloumi, J. Structural, morphological, optical and photodetector properties of sprayed Li-doped ZnO thin films. *J. Mater. Sci.* **2017**, *52*, 10439–10454. [CrossRef]
51. Meziane, K.; El Hichou, A.; El Hamidi, A.; Chhiba, M.; Bourial, A.; Almaggoussi, A. Li concentration dependence of structural properties and optical band gap of Li-doped ZnO films. *Appl. Phys. A* **2017**, *123*, 430. [CrossRef]
52. El-Desoky, M.M.; Ali, M.A.; Afifi, G.; Imam, H. Annealing effects on the structural and optical properties of growth ZnO thin films fabricated by pulsed laser deposition (PLD). *J. Mater. Sci. Mater. Electron.* **2014**, *25*, 5071–5077. [CrossRef]
53. EL-Fadl, A.A.; Mohamad, G.A.; El-Moiz, A.B.A.; Rashad, M. Optical constants of $Zn_{1-x}Li_xO$ films prepared by chemical bath deposition technique. *Phys. B Cond. Matt.* **2005**, *366*, 44–54. [CrossRef]
54. Tezel, F.M.; Kariper, I.A. Structural and Optical Properties of Undoped and Silver, Lithium and Cobalt-doped ZnO thin films. *Surf. Rev. Lett.* **2020**, *27*, 1950138. [CrossRef]
55. Al-Kuhaili, M.F.; Durrani, S.M.A.; El-Said, A.S.; Heller, R. Enhancement of the refractive index of sputtered zinc oxide thin films through doping with Fe_2O_3. *J. Alloys Compd.* **2017**, *690*, 453–460. [CrossRef]
56. Haacke, G. New figure of merit for transparent conductors. *J. Appl. Phys.* **1976**, *47*, 4086–4089. [CrossRef]
57. Badgujar, A.C.; Yadav, B.S.; Jha, G.K.; Dhage, S.R. Room Temperature Sputtered Aluminum-Doped ZnO Thin Film Transparent Electrode for Application in Solar Cells and for Low-Band-Gap Optoelectronic Devices. *ACS Omega* **2022**, *7*, 14203–14210. [CrossRef]
58. Lin, Q.; Zhang, F.; Zhao, N.; Yang, P. Influence of Annealing Temperature on Optical Properties of Sandwiched ZnO/Metal/ZnO Transparent Conductive Thin Films. *Micromachines* **2022**, *13*, 296. [CrossRef]
59. Mahajan, C.M.; Takwale, M.G. Precursor molarity dependent growth rate, microstructural, optical and electrical properties of spray pyrolytically deposited transparent conducting ZnO thin films. *Micro Nanostr.* **2022**, *163*, 107131. [CrossRef]
60. Mallick, A.; Ghosh, S.; Basak, D. Highly conducting and transparent low-E window films with high figure of merit values based on RF sputtered Al and In co-doped ZnO. *Mater. Sci. Semicond. Process.* **2020**, *119*, 105240. [CrossRef]
61. Das, A.; Das, G.; Kabiraj, D.; Basak, D. High conductivity along with high visible light transparency in Al implanted sol-gel ZnO thin film with an elevated figure of merit value as a transparent conducting layer. *J. Alloys Compd.* **2020**, *835*, 155221. [CrossRef]
62. Wu, C.Y.; Chiu, L.C.; Juang, J.Y. High haze Ga and Zr co-doped zinc oxide transparent electrodes for photovoltaic applications. *J. Alloys Compd.* **2022**, *901*, 163678. [CrossRef]

MDPI AG
Grosspeteranlage 5
4052 Basel
Switzerland
Tel.: +41 61 683 77 34

Crystals Editorial Office
E-mail: crystals@mdpi.com
www.mdpi.com/journal/crystals

www.ingramcontent.com/pod-product-compliance
Lightning Source LLC
LaVergne TN
LVHW072252110526
838202LV00106B/2580
* 9 7 8 3 7 2 5 8 2 9 7 1 2 *